The Analysis of Feedback Systems

The Analysis of Feedback Systems

Jan C. Willems

Research Monograph No. 62
THE M.I.T. PRESS
Cambridge, Massachusetts, and London, England

ISBN 0 262 23046 1 (hardcover)

Library of Congress catalog card number: 77-113730

This monograph is dedicated to Margherita.

Contents

Foreword

This is the sixty-second volume in the M.I.T. Research Monograph Series published by the M.I.T. Press. The objective of this series is to contribute to the professional literature a number of significant pieces of research, larger in scope than journal articles but normally less ambitious than finished books. We believe that such studies deserve a wider circulation than can be accomplished by informal channels, and we hope that this form of publication will make them readily accessible to research organizations, libraries, and independent workers.

Howard W. Johnson

Preface

There are two ways, distinct in principle, of mathematically describing physical systems. The first one is called the "input-output description" since it relates external variables. The mathematical model then usually takes the form of an integral equation (the Green's function approach) or more generally, of an operator equation expressing the relationship between the inputs (the variables which can be manipulated) and the outputs (the variables of interest—typically the readings of a set of sensors). Such an input-output description can usually be obtained from some representative experiments. This approach requires minimal knowledge of the physical laws governing the system and of the interconnections within the "black box."

For the "internal description" of a physical system, on the contrary, one uses these physical laws and interconnections as the basis of the mathematical model. This generally takes the form of an ordinary differential equation or a partial differential equation. In the process leading to this model one works with a set of intermediate variables, related to the concept of state. There are thus two parts to mathematical models of internally described systems: a dynamical part—which describes the evolution of the state under the influence of the inputs, and a memoryless part—which relates the output to the state (and sometimes to the instantaneous value of the input as well).

Which of the two descriptions is more convenient depends on the application and the purpose of the analysis or the synthesis that one has in mind. Modern system theory relies heavily on the state formulation for synthesis techniques as exemplified by the highlights of modern control theory: Pontryagin's maximum principle, the regulator problem for linear systems, and the Kalman-Bucy filtering theory.

In the analysis of control systems one usually investigates questions related to stability, continuity, and sensitivity of a closed-loop system. These questions can be treated from both an input-output or a state-space point of view, but it is only very recently that successful results have been obtained and that a sufficiently general framework has been developed to treat them in an input-output setting. The pioneering work in this development has been performed by I. W. Sandberg at Bell Laboratories and G. Zames at NASA-ERC. These authors formulated the stability question (the most important design constraint in feedback control) in an input-output setting. The idea of input-output stability finds its roots in the concept of bounded-input, bounded-output stability and in the work of Nyquist. Nyquist takes the finite integrability of the impulse response as the basic requirement for stability, whereas the concept of bounded-input, bounded-output stability requires that bounded inputs produce bounded outputs. The idea of Nyquist gives an excellent type of stability but unfortunately applies only to a very restricted class of systems, the linear time-invariant systems. The concept of bounded-input, bounded-output stability never has had much success, and very few specific results have been based on it. Moreover, it has been no simple matter to analyze feedback systems in this context, in which they are described by implicit equations. The key in the generalization of these methods to feedback systems has been the introduction of extended spaces. This will be emphasized in the subsequent chapters of this monograph.

This monograph is an attempt to develop further and refine methods based on input-output descriptions for analyzing feedback systems. Contrary to previous work in this area, the treatment heavily emphasizes and exploits the causality of the operators involved. This brings the work into closer contact with the theory of dynamical systems and automata. (In fact, it can be argued that the very definitions of stability and extended spaces are ill-conceived unless the operators involved are explicitly assumed to be causal.)

The monograph is built around Chapter 4, where the relevant concepts of well-posedness, stability, continuity, and sensitivity are introduced. The mathematical foundations for this study will be found in Chapters 2 and 3. In Chapter 2 nonlinear operators are introduced and general conditions for the invertibility or the noninvertibility of nonlinear operators are derived. These conditions rely heavily on the theory of Banach algebras and exploit causality considerations in great detail. Chapter 3 is for the most part devoted to the establishment of a series of inequalities, an unpleasant task that mathematicians usually leave to applied mathematicians and engineers. Inequalities are the workhorses of analysis—and this monograph is no exception to the rule, since these inequalities are essential ingredients for the specific stability and instability criteria described in Chapters 5 and 6. The use of linearization techniques in stability theory is then discussed in Chapter 7.

The monograph is intended primarily for researchers in system theory. The author hopes that it will also be enjoyed by control engineers who are eager to find a unified modern treatment of the analysis of feedback systems, and by mathematicians who appreciate the application of relatively advanced mathematical techniques to engineering questions.

Parts of this monograph appeared in the author's doctoral dissertation entitled "Nonlinear Harmonic Analysis" and submitted in June 1968 to the Department of Electrical Engineering of the Massachusetts Institute of Technology, Cambridge, Massachusetts.

November 1969   Jan Willems

Acknowledgments

The author is glad to have the opportunity to express his indebtedness to Professor Roger W. Brockett of Harvard University, formerly with the Massachusetts Institute of Technology, for his encouragement, moral support, and technical guidance during much of the research reported in this monograph.

Some essential clarifications resulted from discussions with Professor Ronald A. Skoog of the University of California at Berkeley and with Dr. Jon Davis of Queen's University in Kingston, Ontario.

The typing and occasional editing efforts of Miss Karen Erlandson are gratefully acknowledged.

The work reported in this monograph was financially supported in part by the Army Research Office, Durham, under Contract No. DAHCO4-68-C-0026, in part by the National Aeronautics and Space Administration, under Grant No. NGL-22-009-124, and in part by the National Science Foundation under Grant No. GK-14152, all with the Electronic Systems Laboratory of the Massachusetts Institute of Technology.

1 Introduction

1.1 Orientation

A brief outline by chapters of the content of this monograph seems appropriate in order to focus attention on the subject matter, the original results, and the framework of the presentation. The first chapter is introductory; it deals mainly with mathematical preliminaries of a general nature.

The second chapter is devoted to the study of operators and specifically discusses questions related to the invertibility of nonlinear operators. This study is made in an algebraic framework, and special emphasis is placed on the properties of causal (nonanticipatory) operators. Causal operators are indeed of particular interest to engineers and physicists. The concept of causality is roughly equivalent to that of a "dynamical system" and is a basic restriction of physical realizable systems. In the algebraic framework employed in this monograph, causal operators are considered as a subalgebra in the algebra of (in general, nonlinear) operators. Another heavily emphasized and exploited concept is that of extended spaces. These consist of functions which are well-behaved on bounded intervals, but which do not satisfy any regularity conditions at infinity. Extended spaces have not been used extensively in analysis; however, they are the natural setting for the study of causal operators, and they form a very elegant conceptual framework for the study of dynamical systems described, for instance, by an ordinary differential equation with specified initial conditions or by Volterra integral equations.

The analysis in the third chapter is devoted to the derivation of some specific positive operators, which yield the inequalities leading to some general frequency-power formulas and stability conditions.

The basic concepts related to feedback systems are introduced in the fourth chapter. Only the analysis problem is considered, and the main questions investigated are well-posedness, stability, and continuity. This theory is developed in the framework of input-output descriptions of systems, and thus—following the modern trend of mathematical system theory—departs somewhat from the classical methods, which consider undriven systems with initial disturbances.

The fifth chapter discusses the Nyquist criterion and the circle criterion. These yield graphical conditions for stability and instability of linear (possibly time-varying) systems in terms of frequency-response data.

The sixth chapter is devoted to the study of some more complex stability criteria, which apply to systems with a linear time-invariant system in the forward loop and a periodically time-varying gain or a monotone memoryless nonlinearity in the feedback loop.

The final chapter discusses linearization techniques and shows that properly defined linearizations can indeed be successfully used for the analysis of the continuity of feedback systems. This linearization, however, is of a dynamical type and leads to time-varying systems even when the original system is time invariant. The final chapter also contains a simple and rather general class of counterexamples to Aizerman's conjecture.

1.2 Mathematical Preliminaries

This section introduces some notation and definitions which will be freely used throughout this monograph. More details may be found, for instance, in Refs. 1, 2.

A *set* (or *space*) is a collection of objects with a common property. The set, S, of objects with property P is denoted by $S \triangleq \{x \mid x$ has property $P\}$. A *subset* S_1 of a set S, denoted $S_1 \subset S$, is defined as $S_1 = \{x \mid x \in S$ and x has property $P_1\}$ and is sometimes denoted by $S_1 = \{x \in S \mid x$ has property $P_1\}$. The sets R, R^+, I, and I^+ denote respectively the real numbers, the nonnegative real numbers, the integers, and the nonnegative integers.

The *union* of the sets S_1 and S_2, denoted by $S_1 \cup S_2$, is defined as $S_1 \cup S_2 \triangleq \{x \mid x \in S_1 \text{ or } x \in S_2\}$. The *intersection* of the sets S_1 and S_2, denoted by $S_1 \cap S_2$, is defined as $S_1 \cap S_2 \triangleq \{x \mid x \in S_1 \text{ and } x \in S_2\}$. The *Cartesian product* of two sets, denoted by $S_1 \times S_2$, is defined as $S_1 \times S_2 \triangleq \{(x_1,x_2) \mid x_1 \in S_1, x_2 \in S_2\}$.

A *map* F (or *operator* or *function*), from a set S_1 into a set S_2 is a law which associates with every element $x \in S_1$ an element $Fx \in S_2$. S_1 is called the *domain* of the operator. If $S_1' \subset S_1''$ ($S_1'' \subset S_1'$) and if F' and F'' are maps from S_1' into S_2 and S_1'' into S_2 such that $F'x = F''x$ for all $x \in S_1'$ ($x \in S_1''$), then F' is called the *restriction* (an *extension*) of F'' to S_1' (S_1''). A *sequence* is a map from I (I^+) to a set S and will be denoted by $\{x_n\}$, $n \in I$ ($n \in I^+$).

A *metric space* is a set X and a map, d, from $X \times X$ into R^+ such that for all $x, y, z \in X$, the following relations hold: $d(x,y) = d(y,x) \geqslant 0$; $d(x,y) + d(y,z) \geqslant d(x,z)$ (the *triangle inequality*); and $d(x,y) = 0$ if and only if $x = y$.

A sequence $\{x_n\}$, $n \in I^+$, of elements of a metric space X is said to *converge* to a point $x \in X$ if $\lim_{n \to \infty} d(x_n,x) = 0$. This *limit point* x is denoted by $\lim_{n \to \infty} x_n$.

A subset, S, of a metric space, X, is said to be *open* if for any $x \in X$ there exists an $\epsilon > 0$ such that the set $N_\epsilon(x) \triangleq \{y \in X \mid d(x,y) < \epsilon\}$ is a subset of S. A subset, S, of a metric space, X, is said to be *closed* if any converging sequence $\{x_n\}$, $n \in I^+$, of points in S converges to a point in S. A sequence $\{x_n\}$, $n \in I^+$, of elements of a metric space S is said to be a *Cauchy* (or *fundamental*) *sequence* if given any $\epsilon > 0$ there exists an integer N such that $d(x_n,x_m) < \epsilon$ for all $n, m \geq N$. A metric space is said to be *complete* if every Cauchy sequence converges. Completeness is one of the most important properties of metric spaces. A subset of a metric space is said to be *compact* if every bounded sequence has a convergent subsequence.

A *vector space* (sometimes called a *linear space* or a *linear vector space*) is a set V and two maps, one called *addition*, denoted by $+$, from $V \times V$ into V, and the other called *multiplication* from the Cartesian product of the *field* of *scalars* K (which will throughout be taken to be the real or complex number system) and V into V such that for all $x, y, z \in V$ and $\alpha, \beta \in K$:

1. $(x + y) + z = x + (y + z)$;
2. there exists a *zero element*, denoted by 0, with $x + 0 = 0 + x = x$;
3. there exists a *negative element*, denoted by $-x$, with $x + (-x) = 0$ ($y + (-x)$ will be denoted by $y - x$);

4. $x + y = y + z$;
5. $(\alpha + \beta)x = \alpha x + \beta x$;
6. $\alpha(x + y) = \alpha x + \alpha y$;
7. $(\alpha\beta)x = \alpha(\beta x) \triangleq \alpha\beta x$;
8. $1 \cdot x = x$.

A vector space is called a *real* or *complex* vector space according to whether the field K is the real or complex number system. R^n denotes the real vector space formed by the n-tuples of real numbers with addition and multiplication defined in the obvious way. A vector space V is called a *normed vector space* if a map (the *norm*), denoted by $\| \ \|$, from V into R^+ is defined on it, such that:

1. $\|x\| = 0$ if and only if $x = 0$;
2. $\|\alpha x\| = |\alpha| \ \|x\|$;
3. $\|x + y\| \leq \|x\| + \|y\|$ (the *triangle inequality*).

The norm induces a *natural metric* $d(x,y) \triangleq \|x - y\|$, and all statements (e.g. concerning convergence) always refer to this metric, unless otherwise mentioned. Sometimes the norm is subscripted for emphasis, as $\| \ \|_V$, but the subscript will be deleted whenever there is no danger of confusion.

A *Banach space* is a complete normed vector space. This completeness is, of course, to be understood in the topology induced by the natural metric. A very useful class of Banach spaces are the so-called L_p-*spaces*. These consist of Banach space B-valued functions defined on a measurable set $S \subset R$, for which the pth power of the norm is integrable[1] in the case $1 \leq p < \infty$, with the norm defined by

$$\|x\|_{L_p^B(S)} \triangleq \left(\int_S \|x(t)\|_B^p \ dt \right)^{1/p}$$

The space $L_\infty^B(S)$ is defined as the collection of all measurable B-valued functions defined on a measurable set $S \subset R$ which are essentially bounded (i.e., there exists a real number $M < \infty$ such that $\|x(t)\|_B \leq M$ for almost all $t \in S$) with

$$\|x\|_{L_p^B(S)} \triangleq \{\inf M \mid \|x(t)\|_B \leqslant M \text{ almost everywhere on } S\}.$$

---

[1] The integration and measurability considerations refer to Lebesgue measure and integration when the Banach space B is finite-dimensional. Otherwise, these notions are to be interpreted in the sense of Bochner (see, e.g., Ref. 1, p. 78).

The sequence spaces l_p^B are defined in an analogous way, with the integral replaced by a summation. When B is taken to be the real or complex numbers and S is the interval $(-\infty, +\infty)$ for L_p-spaces or $S = I$ for l_p-spaces, then $L_p^B(S)$ and $l_p^B(S)$ will be denoted by L_p and l_p respectively, when no confusion can occur. The L_p-spaces are very often used in analysis. The triangle inequality for L_p-spaces is known as *Minkowski's inequality*. Another useful inequality for L_p-spaces is *Hölder's inequality*, which states that for $f \in L_p(S)$ and $g \in L_q(S)$ with $1/p + 1/q = 1$ and $1 \leqslant p, q \leqslant \infty$, $fg \in L_1(S)$, and

$$\|fg\|_{L_1(S)} \leqslant \|f\|_{L_p(S)} \|g\|_{L_q(S)}.$$

An *inner product space* is a linear vector space, V, with a map, denoted by $\langle \, \rangle$ and called the *inner product*, from $V \times V$ into the scalars K such that for all $x, y, z \in V$ and scalars α, β, the following relations hold:

1. $\langle x,y \rangle = \overline{\langle y,x \rangle}$ (the overbar denotes complex conjugation);
2. $\langle \alpha x + \beta y, z \rangle = \alpha \langle x,z \rangle + \beta \langle y,z \rangle$; and
3. $\langle x,x \rangle \geqslant 0$ and $\langle x,x \rangle = 0$ if and only if $x = 0$.

The inner product induces a *natural norm* $\|x\| = \sqrt{\langle x,x \rangle}$ and all statements (e.g., concerning convergence) always refer to the metric induced by this norm, unless otherwise mentioned. Sometimes the inner product is denoted by $\langle , \rangle_V$, but the subscript will be deleted whenever there is no danger of confusion.

The *Cauchy-Schwartz inequality* states that $|\langle x,y \rangle| \leqslant \|x\| \, \|y\|$.

A *Hilbert space* is a complete inner product space. This completeness is, of course, to be understood in the topology induced by the natural norm. The standard example of a Hilbert space is R^n with $\langle x,y \rangle_{R^n} = \sum_{i=1}^n x_i y_i$ where $x = (x_1, x_2, \ldots, x_n)$ and $y = (y_1, y_2, \ldots, y_n)$. A very useful class of Hilbert spaces are the L_2-*spaces*. These consist of Hilbert space H-valued functions defined on a measurable set $S \subset R$, for which the square of the norm is integrable, and with

$$\langle x,y \rangle_{L_2^H(S)} \triangleq \int_S \langle x(t), y(t) \rangle_H \, dt.$$

Similarly, $l_2^H(S)$ with $S \subset I$ is a Hilbert space, with

$$\langle x,y \rangle_{l_2^H(S)} \triangleq \sum_{n \in S} \langle x_n, y_n \rangle_H.$$

1.3 Transform Theory

Definitions: If $x \in L_1$, then the function X defined by

$$X(j\omega) = \int_{-\infty}^{+\infty} x(t)e^{-j\omega t}\,dt$$

is called the *Fourier transform* of x. Clearly $X \in L_\infty$ and $\|X\|_{L_\infty} \leqslant \|x\|_{L_1}$; if $x(t)$ is real, then $X(j\omega) = \bar{X}(-j\omega)$. Since this transform need not belong to L_1, it is in general impossible to define the inverse Fourier transform. However, if X itself turns out to belong to L_1 then

$$x(t) = \frac{1}{2\pi}\int_{-\infty}^{+\infty} X(j\omega)e^{j\omega t}\,d\omega$$

(As always, this equality is to be taken in the L_1 sense, that is, except on a set of zero Lesbesgue measure.) Thus the need of a slightly more general transform in which the inverse transform can always be defined is apparent. This is done by the limit-in-the-mean transform. It is well known that if x, $y \in L_2 \cap L_1$ then X, $Y \in L_2$ and $\langle x,y \rangle = \langle X,Y \rangle / 2\pi$ (*Parseval's equality*). Let $x \in L_2$. Since $L_1 \cap L_2$ is dense in L_2, i.e., since any L_2-function can be approximated arbitrarily closely (in the L_2 sense) by a function in $L_1 \cap L_2$, there exists a sequence of functions $\{x_n\}$ in $L_2 \cap L_1$ which is a Cauchy sequence and which converges to x (in the L_2 sense). Let X_n be the Fourier transform of x_n. It follows from the Parseval relation that $\|x_n - x_m\| = (2\pi)^{-1/2}\|X_n - X_m\|$ and that $X_n \in L_2$. Thus since L_2 is complete, these transforms, X_n, converge to an element X of L_2. This element X is called the *limit-in-the-mean* transform of x. It follows that the limit-in-the-mean-transform maps L_2 into itself and that $\langle x,y \rangle = \langle X,Y \rangle / 2\pi$ for all x, $y \in L_2$ and their limit-in-the-mean transforms X, Y.

One way of defining a limit-in-the-mean transform is by

$$X(j\omega) = \lim_{T \to \infty} \int_{-T}^{T} x(t)e^{-j\omega t}\,dt$$

where the limit is to be taken in the L_2 sense. (It is easily verified that this induces a particular choice for the Cauchy sequence $\{x_n\}$.) The notation that will be used for the limit-in-the-mean transform is

$$X(j\omega) = \text{l.i.m.}\int_{-\infty}^{+\infty} x(t)e^{-j\omega t}\,dt.$$

With this definition of transforms, the inversion is always possible, and the *inverse transform formula* states that

$$x(t) = \text{l.i.m.} \frac{1}{2\pi} \int_{-\infty}^{+\infty} X(j\omega)e^{j\omega t}\, d\omega.$$

Let $x \in L_2(0,T)$, $T > 0$. Then the sequence $X = \{x_k\}$, $k \in I$, given by

$$x_k = \frac{1}{T} \int_0^T x(t)e^{-jk2\pi t/T}\, dt,$$

is well defined since $L_2(0,T) \subset L_1(0,T)$, and is called the *Fourier series* of $x(t)$. Clearly $X \in l_\infty$ and $x_k = \bar{x}_{-k}$ whenever $x(t)$ is real. The *Parseval relation* states that if x_1, $x_2 \in L_2(0,T)$ and if X_1, X_2 are their Fourier series, then X_1, $X_2 \in l_2$ and $\langle x_1, x_2 \rangle_{L_2(0,T)} = 2\pi \langle X_1, X_2 \rangle_{l_2}$.

In trying to obtain the *inverse Fourier series* formula, the same difficulties are encountered as with the inverse Fourier transform, and the same type of solution is presented. This leads to

$$x(t) = \text{l.i.m.} \sum_{k=-\infty}^{+\infty} x_k e^{jk2\pi t/T}.$$

One way of expressing this l.i.m. summation is by

$$x(t) = \lim_{N \to \infty} \sum_{k=-N}^{N} x_k e^{jk2\pi t/T},$$

where the limit is to be taken in the $L_2(0,T)$ sense.

If $x \in l_1$, then the function X defined by

$$X(z) = \sum_{k=-\infty}^{+\infty} x_k z^{-k}$$

exists for all $|z| = 1$ and is called the *z-transform* of x. In trying to extend this notion to sequences in l_2 the same difficulties and the same solution as in the previous cases present themselves. This leads to the *limit-in-the mean z-transform*

$$X(z) = \text{l.i.m.} \sum_{k=-\infty}^{+\infty} x_k z^{-k}$$

and the *inverse z-transform*

$$x_k = \frac{1}{2\pi} \oint_{|z|=1} X(z)z^{-1}\, dz$$

where the integral is interpreted in the usual manner since

$$L_2(|z| = 1) \subset L_1(|z| = 1).$$

A continuous function, x, from R into K is said to be *almost periodic* if for every $\epsilon > 0$ there exists a real number l such that every interval of the real line of length l contains at least one number τ such that

$$|x(t + \tau) - x(t)| \leqslant \epsilon \text{ for all } t.$$

Some properties of almost periodic functions are:

1. Every almost periodic function is bounded and uniformly continuous.
2. Continuous periodic functions are almost periodic.
3. The sums, products, and limits of uniformly convergent almost periodic functions are almost periodic.
4. The limit, as $T \to \infty$, of the mean value

$$\frac{1}{2T} \int_{-T}^{T} x(t + \tau) \, dt$$

exists, is independent of τ for all almost periodic functions x, and converges uniformly in τ.
5. If x_1 and x_2 are almost periodic functions then so is

$$x_1 * x_2 \triangleq \lim_{T \to \infty} \frac{1}{2T} \int_{-T}^{T} x_1(t - \tau) x_2(\tau) \, d\tau.$$

Moreover, $x_1 * x_2 = x_2 * x_1$ and $x_1 * (x_2 * x_3) = (x_1 * x_2) * x_3$ for all almost periodic functions x_1, x_2, x_3.
6. The function

$$\lim_{T \to \infty} \frac{1}{2T} \int_{-T}^{T} x(t) e^{-j\omega t} \, dt$$

vanishes for all but a countable number of values of ω.
7. The space of almost periodic functions forms an inner product space with

$$\langle x_1, x_2 \rangle = \lim_{T \to \infty} \frac{1}{2T} \int_{-T}^{T} x_1(t) \bar{x}_2(t) \, dt$$

for x_1, x_2 almost periodic functions. (This inner product space is, however, not complete and not separable.) Let x be an almost periodic function and let $\{\omega_k\}$ be the set of values for which the limit in (6) does not vanish and let x_k be the value of that limit for $\omega = \omega_k$. The sequence $\{x_k\}$ is called the *generalized Fourier series* of $x(t)$. If

$x(t)$ is real, then ω belongs to the set $\{\omega_k\}$ if and only if $-\omega$ does and the values x_k associated with ω and $-\omega$ are complex conjugates. The *inverse Fourier series* is defined as

$$x(t) = \lim_{N \to \infty} \sum_{k=-N}^{N} x_k e^{j\omega_k t}$$

This limit, which exists, is to be taken in the metric induced by the inner product on the space of almost periodic functions.[2]

[2] For more details on transform theory, see Refs. 3, 4.

References

1. Hille, E., and Phillips, R. S., *Functional Analysis and Semi-Groups* (second edition), American Mathematical Society, Providence, 1957.
2. Dunford, N., and Schwartz, J. T., *Linear Operators, Parts I and II*, Interscience, New York, 1958.
3. Titchmarsh, E. C., *Introduction to the Theory of Fourier Integrals*, Oxford University Press, Oxford, 1937.
4. Paley, R., and Wiener, N., *Fourier Transforms in the Complex Domain*, American Mathematical Society, Providence, 1934.

2 Nonlinear Operators

2.1 Introduction

Many of the function spaces (e.g., L_p-spaces) encountered in applied mathematics carry the time (or, more generally, a real variable) as an essential parameter. This is particularly true in system theory where the purpose is to study the response of physical systems subject to inputs or initial conditions. Physical systems are furthermore non-anticipatory with respect to the parameter "time" in the sense that past and present values of the outputs do not depend on future values of the inputs. This property leads to the fact that physical systems may be described by a particular class of operators. These operators are referred to as causal operators:[1] they describe a nonanticipatory dependence of outputs on inputs with the range and the domain of the operator parametrized by a real variable, which plays the role of the time.

This causality of operators is very often implicitly used in analysis, but a systematic exploitation of this structure is not often carried out explicitly. This chapter contains a systematic study of extended spaces which are believed to be the natural setting for function spaces parametrized by the time, and of causal operators which describe a non-anticipatory input-output behavior.

Mathematics evolves around equations and their solution. The study

---

[1] Causality is the fundamental property of physically realizable systems. Causality is essentially equivalent to the existence of a state (Ref. 1) and is thus *the* basic property of dynamical systems. For a study of causality from an abstract point of view, see Refs. 2, 3.

of a nonlinear equation usually concentrates on the existence and uniqueness of solutions and on algorithms for the computation of the solution. This chapter contains a number of theorems concerning the solvability of nonlinear operator equations and describes appropriate algorithms for their solution. These involve conditions for the invertibility of operators in terms of contractions, conicity, and positivity considerations. Some of these conditions are standard. Others, which exploit the causality of the operators in an essential way, have, to the author's knowledge, not appeared in the mathematical literature before. All of the theorems which follow have direct applications in the study of feedback systems and are introduced with these applications in mind.

2.2 Operators: Generalities

This section contains a number of general notions concerning nonlinear operators. Most of these notions are standard and can be found in the usual texts on functional analysis.

A mapping from a space X into a space Y is called an *operator* from X into Y. Thus an operator associates with every element $x \in X$ a unique element $y \triangleq Fx \in Y$. X is called the *domain* of F, denoted by Do (F), and Ra $(F) \triangleq \{y \in Y \mid y = Fx, x \in X\} \subset Y$ is called the *range* of F.

An operator F from a metric space X into a metrix space Y is said to be *continuous* if every convergent sequence $\{x_n\}$, $n \in I^+$, yields a convergent sequence $\{Fx_n\}$. It is said to be *Lipschitz continuous* if

$$\sup_{\substack{x,y \in X \\ x \neq y}} \frac{d(Fx, Fy)}{d(x, y)} < \infty.$$

This supremum will be called the *Lipschitz constant* of F on X. An operator F from a normed vector space X into a normed vector space Y is said to be *bounded*[2] if $F0 = 0$ and if

$$\sup_{\substack{x \in X \\ x \neq 0}} \frac{\|Fx\|}{\|x\|} < \infty.$$

This supremum will be called the *bound* of F.

[2] It appears to be no simple matter to define boundedness of a nonlinear operator in a satisfactory way. "Bounded sets into bounded sets" appears to be the most logical—but mathematically least manageable—one. The norm relation $\|Fx\| \leqslant M_1 + M_2 \|x\|$ has been proposed and, although it has some advantages over the one adopted here, it is equally a *compromise* derived from the definition of bounded linear operators. The concept of boundedness adopted here is that of "finite gain." Needless to say, all of the above definitions become equivalent for linear operators.

The operator I from X into itself defined by $Ix \triangleq x$ is called the *identity operator*. The operator O from a vector space X into a vector space Y defined by $Ox \triangleq 0$ is called the *zero operator*.

An operator F from X into Y is said to be *invertible* on X if it is one-to-one on X and onto Y. There then exists an operator, denoted by F^{-1}, from Y into X such that $F^{-1}F = I$ and $FF^{-1} = I$.

Let F_1 and F_2 be operators from X into itself. The operators F_1 and F_2 are said to *commute* on X if $F_1F_2 = F_2F_1$ on X.

Let F_1, F_2 be operators from X into the vector space Y and let α be a scalar. Then $F_1 + F_2$ and αF_1 denote the operators defined for $x \in X$ by $(F_1 + F_2)x \triangleq F_1x + F_2x$ and $(\alpha F_1)x \triangleq \alpha(F_1x)$, respectively.

Let F_1 and F_2 be operators from X into Y and Y into Z, respectively; then F_2F_1 denotes the operator from X into Z defined by $(F_2F_1)x \triangleq F_2(F_1x)$ and is called the *composition* of F_1 and F_2.

Let X and Y be vector spaces over the same field of scalars. An operator L from X into Y is said to be *linear* if for all x_1, $x_2 \in X$ and scalars α, β, the relation $L(\alpha x_1 + \beta x_2) = \alpha Lx_1 + \beta Lx_2$ is satisfied.

Let L be a linear operator from the normed space X into the normed space Y. Then boundedness, continuity, and Lipschitz continuity are equivalent. Moreover, the bound of L equals

$$\sup_{\substack{x \in X \\ \|x\|=1}} \|Lx\|.$$

Let L be an invertible bounded linear operator from a Banach space X onto a Banach space Y. Then L^{-1} is also an invertible bounded linear operator. In fact, linearity of L^{-1} is immediate and boundedness of L^{-1} follows from the closed graph theorem (Ref. 4, p. 47).

Let L be a linear operator from an inner product space X into an inner product space Y. If there exists an operator, L^*, from Y into X such that for all $x \in X$ and $y \in Y$, the relation $\langle y, Lx \rangle_Y = \langle L^*y, x \rangle_X$ holds, then L^* is called the *adjoint* of L. The adjoint L^* is linear and uniquely defined whenever it exists, and $(L^*)^* = L$ (thus $(L^*)^*$ exists if L^* does). By the Riesz representation theorem (Ref. 4, p. 43), L^* exists when L is bounded and X and Y are Hilbert spaces. Moreover, the bound of L^* equals that of L, and L is invertible if and only if L^* is and $(L^{-1})^* = (L^*)^{-1}$. An operator is called *self adjoint* if $L^* = L$.

2.3 Extended Spaces

In this section the notion of extended spaces is introduced. These spaces play an important role in system theory and are very appropriate for the study of causal operators.

Let $S = [T_0, \infty)$ or $(-\infty, +\infty)$.[3] The set S will be referred to as the *time-interval of definition*. Let B be a Banach space, and let $Y(B)$ denote the linear space of B-valued functions defined on S, that is, $Y(B) \triangleq \{x \mid x : S \to B\}$.

Definitions: Let $T \in S$. Then P_T denotes the projection operator on $Y(B)$ defined for $x \in Y(B)$ by

$$(P_T x)(t) \triangleq \begin{cases} x(t) & \text{for } t \leqslant T, t \in S \\ 0 & \text{otherwise} \end{cases}$$

P_T will be called the *truncation operator* and $P_T x$ will be called the *truncation* of x at time T. $\{P_T\}$, $T \in S$, consists thus of a family of projection operators on $Y(B)$.

Let $W \subset Y(B)$ be a Banach space. The *extended space* W_e is defined as $W_e \triangleq \{x \in Y(B) \mid P_T x \in W \text{ for all } T \in S, T \text{ finite}\}$.

The following assumptions[4] are made about the space W:

W.1. The space W is closed under the family of projections $\{P_T\}$, $T \in S$.

W.2. For any $x \in W$, the norm $\|P_T x\|$ is a monotone nondecreasing function of T which satisfies

$$\lim_{T \to \inf S} \|P_T x\| = 0 \quad \text{and} \quad \lim_{T \to \sup S} \|P_T x\| = \|x\|.$$

The family of projection operators $\{P_T\}$, for $T \in S$, is thus assumed to be a resolution of the identity.

W.3. If $x \in W_e$ then $x \in W$ if and only if $\sup_{T \in S} \|P_T x\| < \infty$.

W.4. For any real numbers $t_1, t_2 \in S$, $t_1 \leqslant t_2$, the vector space

$$W_{(t_1, t_2)} \triangleq \{x \in W \mid x(t) = 0 \text{ for } t \notin [t_1, t_2]\}$$

is a closed subspace of W and thus itself a Banach space under the norm of W.

[3] The particular choice of S adopted here does exclude discrete systems, but the adjustments to treat this case are minor and are left to the reader. In fact, S could be taken as an arbitrary subset of the real line. Causality can be defined in terms of arbitrary locally compact Abelian groups. This viewpoint, which follows the development of modern mechanics, is proposed in References 5 and 6.

[4] Not all of these assumptions are necessary for the theorems which follow.

Important spaces which satisfy these assumptions are $L_p^B(S)$ with $1 \leqslant p < \infty$.[5]

The extended space W_e is in general a proper extension of W. For instance, $L_{1e}^B(T_0, \infty)$ is the collection of B-valued measurable functions on $[T_0, \infty)$ whose norm is integrable on *compact* subsets of $[T_0, \infty)$. Note that W_e is a linear space but, very importantly, that it is *not* normed. Notice also that an element $x \in W_e$ belongs to W if and only if the set $M \triangleq \{\alpha \in R \mid \alpha = \|P_T x\|, T \in S\}$ is bounded in R.

2.4 Causal Operators

In this section the notion of a causal operator is formally introduced. It plays a central role in the following chapters and is a fundamental property of physically realizable systems.

Definitions: An operator F from W into itself is said to be *causal* on W if $P_T F$ commutes[6] with P_T on W for all $T \in S$. It is said to be *anticausal* on W if $(I - P_T)F$ commutes with $(I - P_T)$ on W for all $T \in S$. The operator F is said to be *memoryless* on W if F is causal *and* anticausal on W. It is easily seen that a memoryless operator F is necessarily defined, for some map f from $B \times S$ into B, by the relation $(Fx)(t) \triangleq f(x(t), t)$. F is said to be *strongly causal*[7] on W if F is causal on W and if for all $T \in S$ there exists, for any $\epsilon > 0$ and $T' \in S$, $T' \leqslant T$, a real number $\Delta T > 0$ such that for any $x, y \in W$ with $P_{T'} x = P_{T'} y$, the relation $\|P_{T'+\Delta T}(Fx - Fy)\| \leqslant \epsilon \|P_{T'+\Delta T}(x - y)\|$ is satisfied.

Causality is a fundamental property of physically realizable systems. It merely expresses that *past and present output values do not depend on future input values.* Furthermore, many systems encountered in practice contain an integration or a delay in which case they become strongly causal. Strongly causal systems cannot react instantaneously to inputs.

The definition of *causal* and *strongly causal* operators on W_e is completely analogous to these notions on W. A causal operator F from W_e into itself is said to be *memoryless* if for all $T \in S$, the operator $P_T F P_T$ is memoryless on W. If F maps W_e into itself *and* W into itself

---

[5] Indeed, $\{P_T\}$ is not a resolution of the identity if $W = L_\infty$. Most of the theorems which follow, however, still hold.

[6] This definition is equivalent to requiring that $P_T x_1 = P_T x_2$ implies that $P_T F x_1 = P_T F x_2$. Other equivalent definitions are given in Reference 2.

[7] Strong causality is a low-pass filtering condition, which essentially expresses the fact that there is an infinitesimal delay present in the operator.

then F is causal (memoryless) on W if and only if it is causal (memoryless) on W_e. A causal operator F from W into itself has thus a *natural causal extension* to an operator F_e from W_e into itself defined for $x \in W_e$ and any $T \in S$ by $P_T F_e x \triangleq P_T F P_T x$. The above notions of causality are analogously defined for operators which map $W_1(W_{1e})$ into $W_2(W_{2e})$ where W_1 *and* W_2 satisfy assumptions W.1 through W.4 of Section 2.3.

Although extended spaces are not normed it is nevertheless possible to define the notions of continuity and boundedness for causal operators on extended spaces,[8] as is shown by the following definitions.

Let F be a causal operator from W_e into itself. Then F is said to be *locally (Lipschitz) continuous* on W_e if $P_T F P_T = P_T F$ is (Lipschitz) continuous on W for all $T \in S$. It is said to be *locally bounded* on W_e if $F0 = 0$ and if, for all $T \in S$, $P_T F P_T$ is bounded on W.

It is easily established that if F is a Lipschitz-continuous (bounded) operator from W_e into itself, then the Lipschitz constant (bound) K_T of $P_T F P_T$ on W is a monotone nondecreasing function of T.

Let F be a causal operator from W_e into itself. Then F is said to be *Lipschitz continuous (bounded)* on W_e if it is locally Lipschitz continuous (bounded) on W_e and if the Lipschitz constant (bound) of $P_T F P_T$ on W, K_T, satisfies $\sup_{T \in S} K_T = \lim_{T \to \infty} K_T < \infty$. This supremum will be called the *Lipschitz constant (bound)* of F on W_e.

The following theorem relates Lipschitz continuity and boundedness on W_e to the analogous notions on W.

THEOREM 2.1

Let F be a causal operator from W_e into itself, with $F0 = 0$. If F is Lipschitz continuous (bounded) on W_e then F maps W into itself and the Lipschitz constants (bounds) of F on W_e and W are equal. Conversely, if F maps W into itself and is Lipschitz continuous (bounded) on W, then F is Lipschitz continuous (bounded) on W_e and the Lipschitz constants (bounds) of F on W and W_e are equal.

---

[8] It was decided *not* to define boundedness or Lipschitz continuity on extended spaces for operators which are not causal. One could indeed define a bounded operator on W_e as one for which there exists a $K < \infty$ such that $\|P_T F x\| \leqslant K \|P_T x\|$ for all $T \in S$, $x \in W_e$. The author believes, however, that the very existence of extended spaces is only justified when used in connection with causal operators. Indeed, implicit in the definition of extended spaces is the fact that the past and the future are regarded as essentially different; thus unless the operators on those extended spaces show a similar property, such a distinction would appear to be ill founded. The extended space defined in this section is sometimes referred to as the *causal* extension of W.

Proof of Boundedness: Let K and K_e denote the bounds of F on W and W_e respectively. Let $K_e < \infty$ and $x \in W$, then $Fx \in W_e$ and $\|P_T Fx\| \leqslant K\|P_T x\| \leqslant K_e\|x\|$. Thus $Fx \in W$ and $\|Fx\| \leqslant K_e\|x\|$. Hence F maps W into itself, is bounded on W, and $K \leqslant K_e$. Let $x \in W_e$, then $\|P_T Fx\| = \|P_T FP_T x\| \leqslant \|FP_T x\| \leqslant K\|P_T x\|$ which shows that $K_e \leqslant K$. Conversely, let $K < \infty$ and $x \in W_e$, then $\|P_T Fx\| \leqslant K\|P_T x\|$, which shows that F is bounded on W_e and that $K_e \leqslant K$. Let $x \in W$, then $Fx \in W$ and $\|P_T Fx\| \leqslant K_e\|P_T x\| \leqslant K_e\|x\|$. Hence $\|Fx\| \leqslant K_e\|x\|$ and $K \leqslant K_e$.

Proof of Lipschitz continuity: Since by assumption $F0 = 0$, Lipschitz continuity implies boundedness with the Lipschitz constant greater than or equal to the bound. The proof of boundedness thus applies and shows that Lipschitz continuity on W_e implies that F maps W into itself. The rest of the proof evolves parallel to the proof of boundedness, with the inequalities in the differences of $x - y$ and $Fx - Fy$. The details are left to the reader.

Theorem 2.1 thus shows the equivalence for causal operators of Lipschitz continuity (boundedness) on W_e and W respectively. This important property of causal operators plays a central role in stability theory. Theorem 2.1 is easily extended to causal operators which map W_{1e} into W_{2e}.

2.5 Algebras: Generalities

As is well known, the natural setting for the study of linear operators on a Banach space is as a linear Banach algebra. It is not as generally appreciated, however, that most of these ideas carry over to nonlinear operators as well. This is the subject of Sections 2.5 and 2.6. Section 2.5 contains the relevant notions from algebras, and Section 2.6 imbeds nonlinear operators in this algebraic framework.

Let A be a vector space with a mapping (*multiplication*) from $A \times A$ into A defined. Then A is said to be an (*in general nondistributive*) *algebra* if $(xy)z = x(yz) \triangleq xyz$ for all $x, y, z \in A$. An algebra is said to be *left-distributive* if $(x + y)z = xz + yz$ and $(\alpha x)y = \alpha(xy) \triangleq \alpha xy$, for all $x, y, z \in A$ and scalars α. It is said to be *linear* (or *distributive*) if it is left-distributive and if $x(y + z) = xy + xz$ and $x(\alpha y) = \alpha xy$, for all $x, y, z \in A$ and scalars α. A linear algebra is said to be *commutative* if $xy = yx$ for all $x, y \in A$.

An algebra is said to have a *unit* if there exists an element $e \in A$ such that $xe = ex = x$ for all $x \in A$. An element x of an algebra A with a unit e is said to be *regular* (or *invertible*) in A if there exists an element $x^{-1} \in A$ such that $x^{-1}x = xx^{-1} = e$. It is easily seen that there exists at most one unit and one inverse. The unit is always invertible, and $e^{-1} = e$. If x and y are invertible, so is xy, and $(xy)^{-1} = y^{-1}x^{-1}$.

An algebra is said to be a *Banach algebra* if the vector space which defines A is a Banach space, and if $\|xy\| \leqslant \|x\| \|y\|$ for all $x, y \in A$.

A subset A_1 of an algebra A is said to be a *subalgebra* of the algebra A if A_1 is closed under the operations in A (addition, scalar multiplication, and multiplication of elements). A subset A_1 of a Banach algebra A is said to be a *subalgebra* of the Banach algebra A if A_1 is itself a Banach algebra under the norm of A (thus, if A_1 is a subalgebra of the Banach algebra A, then A_1 is closed in the norm topology of A). A subalgebra A_1 of the algebra A is said to be *regular* in A if all elements of A_1 which are invertible in A are invertible in A_1.

A subset M of an algebra A is said to be an *ideal* (*left* or *right* ideal) in A if $xy, yx \in M$ ($xy \in M, yx \in M$) for any $x \in M$ and $y \in A$. Note that if an ideal M in A contains the unit, then $M = A$, and that if it does contain the unit, then no element of M is regular.

Let x be an element of an algebra A with unit e. The *spectrum* of x in A, denoted by $\sigma(x)$, is defined as

$$\sigma(x) \triangleq \{\alpha \text{ complex} \mid \alpha e + x \text{ is not invertible in } A\}.$$

The *resolvent* of x in A, denoted by $\rho(x)$, is the complement in the complex plane of $\sigma(x)$.

Definition: A Banach algebra A with a unit is said to have the *contraction property* if $-1 \notin \sigma(x)$ whenever $\|x\| < 1$. It follows that if a Banach algebra A with a unit has the contraction property then the spectrum of any element is a compact set in the complex plane. Any linear Banach algebra has the contraction property since

$$(e + \alpha)^{-1} = \sum_{n \in I^+} (-\alpha)^n \qquad \text{for} \qquad \|\alpha\| < 1.$$

The following theorem will play an essential role in the method of showing instability of feedback systems that will be introduced in Chapter 4.

THEOREM 2.2

Let A be a left-distributive Banach algebra with a unit e, and let A^+ be a subalgebra of A which contains the unit. Assume that A^+ has the contraction property. Let $x, y \in A^+$, and let C be a connected set in the complex plane. Assume that $x + \alpha y$ is invertible in A for all $\alpha \in C$ and that the set $M \triangleq \{(x + \alpha y)^{-1} \mid \alpha \in C\}$ is a bounded set in A. Then $(x + \alpha y)^{-1} \in A^+$ for all $\alpha \in C$ if and only if $(x + \alpha_0 y)^{-1} \in A^+$ for some $\alpha_0 \in C$.

Proof: Necessity is obvious. For sufficiency it is assumed that $(x + \alpha_0 y)^{-1} \in A^+$ for some $\alpha_0 \in C$. The conclusion clearly holds if $y = 0$. Assume thus that $\|y\| \neq 0$, and let $K \triangleq \sup_{\alpha \in C} \|(x + \alpha y)^{-1}\|$. Notice that $K > 0$. It will first be shown that if $(x + \alpha_1 y)^{-1} \in A^+$ for some complex number $\alpha_1 \in C$, then $x + \alpha y$ is regular in A^+ for all $\alpha \in N \triangleq \{\alpha \text{ complex} \mid |\alpha - \alpha_1| \leqslant \|y\|^{-1} K^{-1}\}$. Write $x + \alpha y$ as $(x + \alpha y) = (e + (\alpha - \alpha_1) y (x + \alpha_1 y)^{-1})(x + \alpha_1 y)$. The claim then follows from the contraction property and the obvious estimate

$$\|(\alpha - \alpha_1) y (x + \alpha_1 y)^{-1}\| \leqslant |\alpha - \alpha_1| \, \|y\| \, \|(x + \alpha_1 y)^{-1}\|.$$

Let P denote the set in the complex plane defined by

$$P \triangleq \{\alpha \text{ complex} \mid x + \alpha y \text{ is regular in } A^+\},$$

and let P^c denote its complement in the complex plane. The theorem claims that $P^c \cap C$ is empty. The proof goes by contradiction: assume that $P^c \cap C$ is not empty. Then

$$d(P \cap C, P^c \cap C) \triangleq \inf_{\substack{\alpha' \in P \cap C \\ \alpha'' \in P^c \cap C}} |\alpha' - \alpha''| \geqslant \|y\|^{-1} K^{-1}.$$

Let

$$N_1 \triangleq \{\alpha \text{ complex} \mid |\alpha - \alpha_1| < \|y\|^{-1} K^{-1}/3, \ \alpha_1 \in P^c \cap C\},$$

$$N_2 \triangleq \{\alpha \text{ complex} \mid |\alpha - \alpha_1| < \|y\|^{-1} K^{-1}/3, \ \alpha_1 \in P \cap C\}.$$

The sets N_1 and N_2 are open, nonempty by assumption, and their union contains C. Hence C is not connected. This contradiction establishes the theorem.

Theorem 2.2 indicates a method for establishing that an inverse belongs to a subalgebra at a value of a parameter by showing invertibility in the subalgebra at *another* value of this parameter and invertibility in the large algebra along a continuous line in the complex plane joining these values of the parameter.

Notice also that the boundedness assumption can be replaced by assuming that C is compact and connected, that A has the contraction property, and that $\|(e + x)^{-1}\| \leqslant (1 - \|x\|)^{-1}$ for $\|x\| < 1$. The assumption implies that $\|(x + \alpha y)^{-1}\|$ is a continuous function of α and is satisfied, for instance, when A is a linear Banach algebra. The boundedness condition then follows from the fact that a continuous real-valued function on a compact set attains its maximum.[9]

2.6 Operator Algebras

The following convention will be assumed throughout:

C.1. All operators that map a vector space into a vector space, map the zero element into the zero element.[10]

It is convenient for purposes of analysis to classify the operators from W_e into itself and those from W into itself. This section contains a listing of some important operator algebras.

Definition: Let $\mathcal{N}(W_e, W_e)$ denote the class of operators from W_e into itself, that is, let $\mathcal{N}(W_e, W_e) \triangleq \{x \mid x\colon W_e \to W_e\}$. Let addition and scalar multiplication in $\mathcal{N}(W_e, W_e)$ be defined in the obvious way, and let multiplication be defined as composition of maps. These definitions lead immediately to Theorem 2.3.

THEOREM 2.3
The operators in $\mathcal{N}(W_e, W_e)$ form a left-distributive algebra with a unit.

Some important subalgebras of $\mathcal{N}(W_e, W_e)$ are:

1. The class $\mathcal{N}^+(W_e, W_e)$ of causal operators from W_e into itself.
2. The class $\mathcal{L}(W_e, W_e)$ of linear operators from W_e into itself.
3. The locally continuous operators from W_e into itself.
4. The locally Lipschitz continuous operators from W_e into itself.
5. The locally bounded operators from W_e into itself.
6. The Lipschitz continuous operators from W_e into itself.
7. The bounded operators from W_e into itself.
8. The memoryless operators from W_e into itself.

[9] For more details on algebras, see Reference 4.

[10] It is possible to dispense with this assumption at the expense of a somewhat laborious notational framework. Any operator, however, can be fixed up so as to satisfy this condition by defining $F'x \triangleq Fx - F0$, and replacing Fx by $F'x + F0$.

    Subalgebras 1–8 and any intersection of them form subalgebras of $\mathcal{N}(W_e,W_e)$ which contain the unit.

9. The strongly causal operators from W_e into itself form a subalgebra of $\mathcal{N}(W_e,W_e)$ which does not contain the unit. This class of operators is in fact a subalgebra of $\mathcal{N}^+(W_e,W_e)$ and an ideal in the class of locally Lipschitz continuous operators from W_e into itself. Thus no strongly causal operator is invertible in the class of locally Lipschitz continuous causal operators.

10. Let the time-interval of definition S be $(-\infty,+\infty)$. Let τ be a real number. Then the *translation operator*, denoted by T_τ, is defined for $x \in W_e$ by $(T_\tau x)(t) \triangleq x(t + \tau)$. An operator F from W_e into itself is said to be *time invariant* on W_e if for any $\tau \in R$, T_τ and F commute on W_e. Let the time-interval of definition be $[T_0,\infty)$. Let $\tau \leqslant 0$ be a real number. Then the translation operator, denoted by T_τ, is defined for $x \in W_e$ by $(T_\tau x)(t) \triangleq x(t + \tau)$ for $t + \tau \geqslant T_0$, and 0 otherwise. A causal[11] operator F from W_e into itself is said to be *time invariant* on W_e if T_τ and F commute on W_e for all $\tau \leqslant 0$. The time-invariant operators form a subalgebra of $\mathcal{N}(W_e,W_e)$ which contains the unit.

    It is easy to verify that $\mathcal{L}(W_e,W_e)$ is a regular subalgebra of $\mathcal{N}(W_e,W_e)$. If $S = (-\infty,+\infty)$ then the time-invariant operators form a regular subalgebra of $\mathcal{N}(W_e,W_e)$. If $S = [T_0,\infty)$, then the time-invariant operators form a regular subalgebra of $\mathcal{N}^+(W_e,W_e)$ and $\mathcal{N}(W_e,W_e)$. The memoryless operators also form a regular subalgebra of $\mathcal{N}^+(W_e,W_e)$ and $\mathcal{N}(W_e,W_e)$.

    In the same way as for the algebra of operators from W_e into itself, one can consider the algebra of operators from W into itself. Thus, let $\mathcal{N}(W,W)$ denote the class of operators from W into itself, i.e., $\mathcal{N}(W,W) \triangleq \{x \mid x\colon W \to W\}$. Let addition, scalar multiplication, and multiplication of elements be defined in the obvious way. Then there follows:

THEOREM 2.4

    The operators in $\mathcal{N}(W,W)$ form a left-distributive algebra with a unit.

---

[11] Notice that time invariance is only defined for causal operators when $S = [T_0,\infty)$. An alternative approach would be to define time invariance when $S = [T_0,\infty)$ as the property of an operator with a time-invariant backwards extension to $S = (-\infty,+\infty)$ (see Section 2.9).

The analogous subalgebras thus become:

1. The class $\mathcal{N}^+(W,W)$ of causal operators from W into itself.
   The class $\mathcal{N}^-(W,W)$ of anticausal operators from W into itself.
   The class $\mathcal{N}^\circ(W,W)$ of memoryless operators from W into itself.
2. The class $\mathcal{L}(W,W)$ of bounded linear operators from W into itself.
3. The continuous operators from W into itself.
4. The Lipschitz continuous operators from W into itself.
5. The bounded operators from W into itself.
6. The time-invariant operators from W into itself (*time invariance* is defined completely analogously to that notion in W_e).
   Subalgebras 1–6 and any intersection of them contain the unit.
7. The strongly causal operators from W_e into itself form a subalgebra of $\mathcal{N}(W,W)$ which does not contain the unit. This class of operators is in fact a subalgebra of $\mathcal{N}^+(W,W)$ and an ideal in the class of Lipschitz continuous operators from W into itself. Thus, no strongly causal operator is invertible in the class of Lipschitz continuous operators from W into itself.

$\mathcal{L}(W,W)$, $\mathcal{N}^\circ(W,W)$, and the time-invariant operators with $S = (-\infty, +\infty)$ are regular subalgebras of $\mathcal{N}(W,W)$. The algebra $\mathcal{N}^\circ(W,W)$ and the time-invariant operators with $S = [T_0, \infty)$ are regular subalgebras of $\mathcal{N}^+(W,W)$ and $\mathcal{N}(W,W)$.

For the purposes of analysis it is often mandatory to work with Banach algebras rather than simply algebras. Operators which belong to Banach algebras can be "measured" and this leads to the possibility of establishing invertibility conditions for operators. Two important norms on the space of operators will be considered: the first one takes the Lipschitz constant of an operator as its measure and the second one takes the bound of an operator as its measure.

Definition: Let $\tilde{\mathcal{B}}(W,W)$ denote the space of Lipschitz continuous operators from W into itself which map zero into itself, and define the norm of an element $F \in \tilde{\mathcal{B}}(W,W)$ by

$$\|F\|_\Delta \triangleq \sup_{\substack{x,y \in W \\ x \neq y}} \frac{\|Fx - Fy\|}{\|x - y\|}.$$

Let addition, scalar multiplication, and multiplication of elements be defined in $\tilde{\mathcal{B}}(W,W)$ in the usual way.

THEOREM 2.5

The algebra $\tilde{\mathcal{B}}(W,W)$ is a left-distributive Banach algebra with a unit.

Proof: The only element in the proof which is not immediate is the statement that $\tilde{\mathscr{B}}(W,W)$ is a Banach space. Let $\{F_n\}$, $n \in I^+$, be a Cauchy sequence in $\tilde{\mathscr{B}}(W,W)$. Then $\{F_n x\}$, for any $x \in W$, is a Cauchy sequence in W. Let $Fx \triangleq \lim_{n\to\infty} F_n x$. Clearly F maps W into itself and $F0 = 0$. Since

$$\|Fx - Fy\| = \lim_{n\to\infty} \|F_n x - F_n y\| \leqslant \sup_{n\in I^+} \|F_n\|_\Delta \|x - y\|,$$

$$\sup_{\substack{x,y\in W \\ x\neq y}} \frac{\|Fx - Fy\|}{\|x - y\|} < \infty,$$

and $F \in \tilde{\mathscr{B}}(W,W)$. It remains to be shown that $\|F_n - F\|_\Delta \to 0$. However, since for $x, y \in W$, $x \neq y$,

$$\frac{\|F_n x - F_n y - (Fx - Fy)\|}{\|x - y\|}$$

$$= \frac{\|F_n x - F_n y - \lim_{m\to\infty}(F_m x - F_m y)\|}{\|x - y\|}$$

it follows that $\|F_n - F\|_\Delta \leqslant \sup_{m\geqslant n} \|F_n - F_m\|_\Delta$, which yields the convergence since $\{F_n\}$ is Cauchy in $\tilde{\mathscr{B}}(W,W)$.

Definition: Let $\mathscr{B}(W,W)$ denote the space of operators from W into itself which map zero into itself and define the norm of an element $F \in \mathscr{B}(W,W)$ by

$$\|F\| \triangleq \sup_{\substack{x\in W \\ x\neq 0}} \frac{\|Fx\|}{\|x\|}.$$

Let addition, scalar multiplication, and multiplication of elements be defined in $\mathscr{B}(W,W)$ in the usual way.

THEOREM 2.6
   The algebra $\mathscr{B}(W,W)$ is a left-distributive Banach algebra with a unit.

Proof: The proof requires only minor modifications from the proof of the previous theorem and is left to the reader.

   Important subalgebras of $\tilde{\mathscr{B}}(W,W)$ include:

1. $\tilde{\mathscr{B}}^+(W,W) \triangleq \mathscr{N}^+(W,W) \cap \tilde{\mathscr{B}}(W,W)$,
   $\tilde{\mathscr{B}}^-(W,W) \triangleq \mathscr{N}^-(W,W) \cap \tilde{\mathscr{B}}(W,W)$,
   $\tilde{\mathscr{B}}^\circ(W,W) \triangleq \mathscr{N}^\circ(W,W) \cap \tilde{\mathscr{B}}(W,W)$;

2. $\mathscr{L}(W,W)$; and
3. the time-invariant operators in $\tilde{\mathscr{B}}(W,W)$.

The classes 1–3, or any intersection of them, form subalgebras of $\tilde{\mathscr{B}}(W,W)$ that contain the unit.

Important subalgebras of $\mathscr{B}(W,W)$ include:

1. $\mathscr{B}^+(W,W) \triangleq \mathscr{N}^+(W,W) \cap \mathscr{B}(W,W)$,
   $\mathscr{B}^-(W,W) \triangleq \mathscr{N}^-(W,W) \cap \mathscr{B}(W,W)$,
   $\mathscr{B}^\circ(W,W) \triangleq \mathscr{N}^\circ(W,W) \cap \mathscr{B}(W,W)$;
2. $\mathscr{L}(W,W)$; and
3. the time-invariant operators in $\mathscr{B}(W,W)$.

The classes 1–3, or any intersection of them, form subalgebras of $\mathscr{B}(W,W)$ which contain the unit.

The only difficulty in showing that the above claims are correct occurs in the demonstration of closedness of the subspaces. This, however, follows readily by contradiction. The linear and the time-invariant operators again form regular subalgebras.

Theorems 2.5 and 2.6 thus succeed in putting a Banach algebra structure on a large subclass of operators in $\mathscr{N}(W,W)$. This result depends directly on the fact that W itself is a Banach space, and hence cannot immediately be generalized to operators in $\mathscr{N}(W_e,W_e)$, since W_e is not normed. There are, however, more restricted subclasses of $\mathscr{N}(W_e,W_e)$ where such a possibility exists.

Definitions: Let $\tilde{\mathscr{B}}^+(W_e,W_e)$ denote the space of Lipschitz continuous causal operators from W_e into itself which map zero into itself, and define the norm of an element $F \in \tilde{\mathscr{B}}^+(W_e,W_e)$ by

$$\|F\|_\Delta \triangleq \sup_{\substack{x,y \in W_e \\ T \in S \\ P_T x \neq P_T y}} \frac{\|P_T F x - P_T F y\|}{\|P_T x - P_T y\|}$$

Let $\mathscr{B}^+(W_e,W_e)$ denote the space of Lipschitz continuous causal operators from W_e into itself which map zero into itself, and define the norm of an element $F \in \mathscr{B}^+(W_e,W_e)$ by

$$\|F\| \triangleq \sup_{\substack{x \in W_e \\ T \in S \\ P_T x \neq 0}} \frac{\|P_T F x\|}{\|P_T x\|}$$

Let addition, scalar multiplication, and multiplication of elements in $\tilde{\mathscr{B}}^+(W_e,W_e)$ and $\mathscr{B}^+(W_e,W_e)$ be defined in the usual way.

THEOREM 2.7

The algebras $\tilde{\mathscr{B}}^+(W_e,W_e)$ and $\mathscr{B}^+(W_e,W_e)$ are left-distributive Banach algebras with a unit.

Proof: Since by Theorem 2.1, $\tilde{\mathscr{B}}^+(W_e,W_e)$ and $\mathscr{B}^+(W_e,W_e)$ are isometrically isomorphic to respectively $\tilde{\mathscr{B}}^+(W,W)$ and $\mathscr{B}^+(W,W)$, the theorem follows from Theorems 2.5 and 2.6.

Thus, as was shown in Theorem 2.1 and pointed out in the proof of Theorem 2.7, $\tilde{\mathscr{B}}^+(W_e,W_e)$ is isometrically isomorphic to $\tilde{\mathscr{B}}^+(W,W)$, and $\mathscr{B}^+(W_e,W_e)$ is isometrically isomorphic to $\mathscr{B}^+(W,W)$. As a consequence, the identical notation for the norms on these spaces is, albeit abusive, not a source of confusion.

2.7 Contractions, Conic Operators, Positive Operators

This section contains a number of concepts around which invertibility theorems will be established. Recall the assumption that for all operators $F0 = 0$.

Let $F \in \tilde{\mathscr{B}}(W,W)$. Then F is said to be a *contraction* on W if $\|F\|_\Delta < 1$. Let $F \in \tilde{\mathscr{B}}^+(W_e,W_e)$. Then F is said to be a *contraction* on W_e if $\|F\|_\Delta < 1$. Notice that by Theorem 2.1 the causal contractions on W stand in one-to-one relation to the contractions on W_e. Let $F \in \mathscr{B}(W,W)$. Then F is said to be *attenuating* on W if $\|F\| < 1$. Let $F \in \mathscr{B}^+(W_e,W_e)$. Then F is said to be *attenuating* on W_e if $\|F\| < 1$. Note that by Theorem 2.1 the causal attenuating operators on W stand in one-to-one relation with the attenuating operators on W_e.

Definitions: Let r be a nonnegative real number and let c be a scalar. Then $F \in \tilde{\mathscr{B}}(W,W)$ is said to be *incrementally (strictly) interior conic* on W with *center* c and *radius* r if $\|F - cI\|_\Delta \leqslant r(<r)$. Let $F \in \mathscr{N}(W,W)$. Then F is said to be *incrementally (strictly) exterior conic* on W with *center* c and *radius* r if for all $x, y \in W$, $\|(F - cI)x - (F - cI)y\| \geqslant r\|x - y\|$ (for some $\epsilon > 0$, $\geqslant (r + \epsilon)\|x - y\|$). Let $F \in \mathscr{B}(W,W)$. Then F is said to be *(strictly) interior conic* on W with *center* c and *radius* r if $\|F - cI\| \leqslant r(<r)$. Let $F \in \mathscr{N}(W,W)$. Then F is said to be *(strictly) exterior conic* on W with *center* c and *radius* r if for all $x \in W$, $\|(F - cI)x\| \geqslant r\|x\|$ (for some $\epsilon > 0$, $\geqslant (r + \epsilon)\|x\|$).

The analogous notions can also be defined on W_e. Thus, let r be a nonnegative real number and let c be a scalar. Then $F \in \mathscr{N}^+(W_e,W_e)$ with $F0 = 0$ is said to be *incrementally interior (strictly interior, exterior, strictly exterior), conic* on W with *center c* and *radius r* if for all $x, y \in W_e$ and all $T \in S$, $\|P_T(F - cI)x - P_T(F - cI)y\| \leqslant r \|P_T(x - y)\|$ (for some $\epsilon > 0$, $\leqslant (r - \epsilon) \|P_T(x - y)\|$, $\geqslant r \|P_T(x - y)\|$, $\geqslant (r + \epsilon) \|P_T(x - y)\|$). It is said to be *interior (strictly interior, exterior, strictly exterior), conic* on W_e with *center c* and *radius r* if for all $x \in W_e$ and $T \in S$, $\|P_T(F - cI)x\| \leqslant r \|P_T x\|$ (for some $\epsilon > 0$, $\leqslant (r - \epsilon) \|P_T x\|$, $\geqslant r \|P_T x\|$, $\geqslant (r + \epsilon) \|P_T x\|$).

Conicity on W_e thus essentially refers to conicity on W of the operators $P_T F P_T$ for all $T \in S$. In many cases of interest the space W which is assumed to be a Banach space is actually a Hilbert space. The case in which W is a Hilbert space is therefore studied more intensively. The following concepts will be useful for this purpose.

Definitions: Let W be a Hilbert space, and let a and b be scalars. Then $F \in \mathscr{N}(W,W)$ is said to be *incrementally inside (outside) the sector* $[a,b]$ on W if for all $x, y \in W$,

$$\text{Re} \langle (F - aI)x - (F - aI)y, (F - bI)x - (F - bI)y \rangle \leqslant 0 \quad (\geqslant 0).$$

It is said to be *incrementally strictly inside (outside) the sector* $[a,b]$ on W if for some $\epsilon > 0$, and all $x, y \in W$

$$\text{Re} \langle (F - aI)x - (F - aI)y, (F - bI)x - (F - bI)y \rangle$$
$$\leqslant -\epsilon \|x - y\|^2 \quad (\geqslant \epsilon \|x - y\|^2).$$

It is said to be *inside (outside) the sector* $[a,b]$ on W if for all $x \in W$,

$$\text{Re} \langle (F - aI)x, (F - bI)x \rangle \leqslant 0 \quad (\geqslant 0).$$

It is said to be *strictly inside (outside) the sector* $[a,b]$ on W if for all $x \in W$,

$$\text{Re} \langle (F - aI)x, (F - bI)x \rangle \leqslant -\epsilon \|x\|^2 \quad (\geqslant \epsilon \|x\|^2).$$

The analogous notions on W_e are defined as follows: Let W be a Hilbert space. Let a and b be scalars. Then $F \in \mathscr{N}^+(W_e,W_e)$ is said to be *incrementally inside (strictly inside, outside, strictly outside) the sector* $[a,b]$ on W_e if for all $x, y \in W_e$ and $T \in S$,

$$\text{Re} \langle P_T(F - aI)x - P_T(F - aI)y, P_T(F - bI)x - P_T(F - bI)y \rangle$$
$$\leqslant 0 \text{ (for some } \epsilon > 0, \quad \leqslant - \epsilon \|P_T(x - y)\|^2,$$
$$\geqslant 0, \quad \geqslant \epsilon \|P_T(x - y)\|^2).$$

It is said to be *inside* (*strictly inside, outside, strictly outside*) *the sector* [*a,b*] on W_e if for all $x \in W_e$ and $T \in S$

$$\text{Re } \langle P_T(F - aI)x, P_T(F - bI)x \rangle \leqslant 0$$

$$(\text{for some } \epsilon > 0, \quad \leqslant -\epsilon \|P_T x\|^2, \quad \geqslant 0, \quad \geqslant \epsilon \|P_T x\|^2).$$

It is clear that contractive operators are particular cases of conic operators (with center 0 and radius 1). Furthermore, interior conicity on W_e and W are by Theorem 2.1 equivalent notions for causal operators. This, however, is *not* the case for exterior conicity.[12]

For operators on a Hilbert space, it is in general easier to verify sector conditions than conicity conditions. The following theorem establishes a relationship between conic operators and sector operators.

THEOREM 2.8

Let W be a Hilbert space and let $F \in \mathcal{N}(W,W)$ (or $F \in \mathcal{N}^+(W_e, W_e)$). Then F is (incrementally) (strictly) interior (exterior) conic on W (or W_e) with center c and radius $|r|$ if and only if F is (incrementally) (strictly) inside (outside) the sector [*a,b*] on W (or W_e) with $a = c - r$ and $b = c + r$.

Proof: Since

$$\text{Re } \langle (F - aI)x, (F - bI)x \rangle = \text{Re } \langle (F - cI + rI)x, (F - cI - rI)x \rangle$$

$$= \|(F - cI)x\|^2 - |r|^2 \|x\|^2,$$

the result follows.

It follows from Theorems 2.1 and 2.8 that the operators that are inside a sector on W_e stand in one-to-one correspondence with the causal operators which are inside that sector on W. This correspondence again does not exist for causal operators which are *outside* a sector on W.

An important role will be played in the sequel by *positive* operators. They are generalizations of nonnegative definite linear operators on Hilbert spaces and have been referred to as *dissipative, passive,* or *monotone* operators.[13]

Definitions: If W is a Hilbert space and $F \in \mathcal{N}(W,W)$, then F is said to be *incrementally positive* on W if $\text{Re } \langle x - y, Fx - Fy \rangle \geqslant 0$ for all

---

[12] It suffices therefore to consider a time-delay on $L_2(-\infty, +\infty)$. This operator is exterior conic with center 0 and radius 1 on $L_2(-\infty, +\infty)$, but *not* on $L_{2e}(-\infty, +\infty)$.

[13] An important reference for the implications of positivity to well-posedness problems for partial differential equations is Reference 7.

$x, y \in W$. It is said to be *positive* on W if Re $\langle x, Fx \rangle \geqslant 0$ for all $x \in W$. It is said to be *(incrementally) strictly positive* on W if for some $\epsilon > 0$, the difference $F - \epsilon I$ is *(incrementally) positive on W*. F is said to be *(incrementally) (strictly) negative* on W if the operator $-F$ is (incrementally) (strictly) negative on W. Let $F \in \mathcal{N}^+(W_e, W_e)$. Then F is said to be *positive* on W_e if Re $\langle P_T x, P_T Fx \rangle \geqslant 0$ for all $x \in W_e$ and $T \in S$. The notions of negativity, incremental positivity (negativity), and strict (incremental) positivity (negativity) on W_e are defined in obvious analogy with those notions on W.

The relationship between positive operators on W and W_e is treated in the following theorem.

THEOREM 2.9

Let $F \in \mathcal{N}^+(W_e, W_e)$ and $\mathcal{N}^+(W, W)$. Then F is (incrementally) (strictly) positive (negative) on W_e if and only if F is (incrementally) (strictly) positive (negative) on W.

Proof: Since P_T is a projection onto a closed subspace of W for all $T \in S$, and F is causal, it follows that for all $x \in W$,

$$\langle P_T x, P_T Fx \rangle = \langle P_T x, P_T F P_T x \rangle = \langle P_T x, F P_T x \rangle;$$

this shows that positivity on W indeed implies positivity on W_e. Conversely, assume that F is positive on W_e but that for some $x \in W$, $\langle x, Fx \rangle < 0$. Since $\lim_{T \to \infty} P_T x = x$, this implies that $\langle P_T x, Fx \rangle = \langle P_T x, P_T Fx \rangle < 0$ for some $T \in S$. This contradiction ends the proof of the theorem.

Remark: It is sometimes easier to perform certain calculations with $S = (-\infty, +\infty)$ and then to draw conclusions for $S = [T_0, \infty)$. The following considerations may be helpful in this regard. Let $S = (-\infty, +\infty)$, let W be given (over S), and let F be an operator from W into itself. Let $S' = [T_0, \infty)$ and let W' be the space having S' as the interval of definition and consisting of those B-valued functions on S' that when extended by zero on $S - S'$ yield elements on W. Let $\|x\|_{W'}$ be equal to the norm of the element of W that when projected on the subspace of functions with support on S' yields $x(t)$ as the value of x for $t \in S'$. Let $x' \in W'$ and $x \in W$ be defined by

$$x(t) = \begin{cases} x'(t) & \text{for} \quad t \in S' \\ 0 & \text{otherwise} \end{cases}$$

Let $F': W' \to W'$ be defined as $(F'x')(t) = (Fx)(t)$ for $t \in S'$. It is clear that F' is well defined.

The space W' and the operator F' will be called a *restriction* of W and F to $S' = [T_0, \infty)$. The same procedure allows one to obtain a restriction on W'_a of an operator on W_a.

The restriction of an operator preserves some of the essential properties of the original operator. These include, for instance, linearity, causality, and time invariance. It is easily verified that the restriction of an operator is (incrementally) positive if the original operator is itself (incrementally) positive. The same holds for interior (incremental) conic, contraction, and attenuating operators. Exterior conicity, however, is in general not preserved under this restriction unless the operator also happens to be causal.

2.8 Conditions for Invertibility of Nonlinear Operators

2.8.1 General Invertibility Conditions on W Involving Contractions, Conicity, Sector Conditions, and Positivity

This section contains general conditions for the invertibility of nonlinear operators on W. It should be remarked that at no point will the special structure of W introduced in Section 2.3 be exploited. The invertibility conditions involve contractions, conicity, sector conditions, and positivity. All of these conditions are immediately related to the celebrated contraction mapping principle[14] and the conditions involving contractions and positivity have previously appeared in the mathematical literature. The conditions involving conicity and sector conditions are new,[15] and should, as sufficient conditions for invertibility of nonlinear operators, be of considerable interest in applied mathematics.

The conditions of some of the theorems are on occasion rather involved, and it is therefore suggested that the reader concentrate first on Theorem 2.11, Theorem 2.12, Corollary 2.14.1, and Theorem 2.15.

It is worthwhile to mention that all the invertibility theorems which follow are in essence based on the contraction mapping principle. They thus yield as an important side aspect a convergent recursive algorithm for evaluating the inverse. In these invertibility theorems special

[14] See Reference 8, p. 34. For some folklore about fixed-point theorems, see Reference 9.

[15] Related, but much more restrictive, conditions have appeared in the mathematical literature (Ref. 8, p. 296). The author draws his inspiration from the work of Zames (Ref. 10), who proves stability theorems using similar conditions. After the link between stability and invertibility is established (as will be done in Chapter 4) these invertibility theorems become apparent. Their derivation is of independent interest, however.

emphasis is placed on whether or not a particular inverse belongs to the subalgebras to which the original operator belongs. The interest in this aspect stems mainly from the applications to instability conditions as will become apparent in Chapter 4.

Consider the Banach algebra $\tilde{\mathscr{B}}(W,W)$ introduced in Section 2.5. Recall that $\tilde{\mathscr{B}}^{+}(W,W)$ is isometrically isomorphic to $\tilde{\mathscr{B}}^{+}(W_e,W_e)$ and that invertibility conditions on $\tilde{\mathscr{B}}^{+}(W_e,W_e)$ thus lead to identical invertibility conditions on $\tilde{\mathscr{B}}^{+}(W_e,W_e)$.

THEOREM 2.10

The algebra $\tilde{\mathscr{B}}(W,W)$ and any of its subalgebras which contain the unit have the contraction property.

Proof: Let $F \in \tilde{\mathscr{B}}(W,W)$ with $\|F\|_\Delta < 1$. Then by the contraction mapping principle, $I + F$ is invertible on W and its inverse is Lipschitz continuous. In fact,

$$\|(I + F)^{-1}\|_\Delta \leqslant \frac{1}{1 - \|F\|_\Delta},$$

and the solution to the equation $x = -Fx + u$ with $u \in W$ given and $x \in W$ unknown can be solved by successive approximations with $x_{n+1} = -Fx_n + u$, $n \in I^+$. The resulting sequence $\{x_n\}$, with $n \in I^+$, is a Cauchy sequence in W and converges for any choice of $x_0 \in W$ to the unique solution $(I + F)^{-1}u$. This algorithm for computing the solution shows that any subalgebra of $\tilde{\mathscr{B}}(W,W)$ also has the contraction property.

Theorem 2.10 immediately leads to the invertibility condition of Theorem 2.11.

THEOREM 2.11

Let $F \in \mathcal{N}(W,W)$ be a contraction on W. Then $-1 \notin \sigma(F)$ in $\tilde{\mathscr{B}}(W,W)$. Moreover, if F belongs to a subalgebra of $\tilde{\mathscr{B}}(W,W)$ that contains the unit, then so does $(I + F)^{-1}$. Finally,

$$\|(I + F)^{-1}\|_\Delta \leqslant \frac{1}{1 - \|F\|_\Delta},$$

THEOREM 2.12

Let $F_1, F_2 \in \mathcal{N}(W,W)$. Then $-1 \notin \sigma(F_1F_2)$ in $\mathcal{N}(W,W)$ if there exists a scalar c such that $-1 \notin \sigma(cF_2)$ in $\mathcal{N}(W,W)$ and $(F_1 - cI)$ $F_2(I + cF_2)^{-1}$ is a contraction on W. Moreover, $(I + F_1F_2)^{-1} \in \tilde{\mathscr{B}}(W,W)$

if $(I + cF_2)^{-1} \in \tilde{\mathscr{B}}(W,W)$, and if F_1 and F_2 belong to a subalgebra of $\tilde{\mathscr{B}}(W,W)$ that contains the unit, then $(I + F_1F_2)^{-1}$ belongs to that subalgebra if and only if $(I + cF_2)^{-1}$ does. Finally,

$$\|(I + F_1F_2)^{-1}\|_\Delta \leqslant \frac{\|(I + cF_2)^{-1}\|_\Delta}{1 - \|(F_1 - cI)F_2(I + cF_2)^{-1}\|_\Delta}$$

Proof: Since $I + F_1F_2 = (I + (F_1 - cI)F_2(I + cF_2)^{-1})(I + cF_2)$, the operator $I + F_1F_2$ is the product of two elements of $\mathscr{N}(W,W)$ both of which are invertible in $\mathscr{N}(W,W)$, and is thus itself invertible in $\mathscr{N}(W,W)$. To prove the second part of the theorem, express $I + F_1F_2$ as $(I + cF_2) + (F_1 - cI)F_2$. Clearly, $(I + cF_2) + \alpha(F_1 - cI)F_2$ is invertible in $\tilde{\mathscr{B}}(W,W)$ for all $0 \leqslant \alpha \leqslant 1$ and the inverse for $\alpha = 0$ is $(I + cF_2)^{-1}$. It thus follows from Theorem 2.2 that the inverse for $\alpha = 1$, $(I + F_1F_2)^{-1}$, belongs to exactly those subalgebras to which $(I + cF_2)^{-1}$ belongs.

COROLLARY 2.12.1
    Let F_1, $F_2 \in \mathscr{N}(W,W)$. Then $-1 \notin \sigma(F_1F_2)$ in $\mathscr{N}(W,W)$ if for some scalar c and $r > 0$, F_1 is incrementally strictly interior conic on W with center c and radius r, $-1 \notin \sigma(cF_2)$ in $\mathscr{N}(W,W)$, and F_2 satisfies for all x, $y \in W$ the inequality $\|(I + cF_2)x - (I + cF_2)y\| \geqslant r\|F_2x - F_2y\|$. Moreover, $(I + F_1F_2)^{-1} \in \tilde{\mathscr{B}}(W,W)$ if $(I + cF_1)^{-1} \in \tilde{\mathscr{B}}(W,W)$, and if F_1 and F_2 belong to a subalgebra of $\tilde{\mathscr{B}}(W,W)$ that contains the unit then $(I + F_1F_2)^{-1}$ belongs to that subalgebra if and only if $(I + cF_2)^{-1}$ does. Finally,

$$\|(I + F_1F_2)^{-1}\|_\Delta \leqslant \frac{\|(I + cF_2)^{-1}\|_\Delta}{1 - \|(F_1 - cI)\|_\Delta/r} \leqslant \frac{1 + |c|/r}{1 - \|F_1 - cI\|_\Delta/r}$$

Proof: The conditions of the Corollary imply that $\|F_2(I + cF_2)^{-1}\|_\Delta \leqslant r^{-1}$ which yields $\|(F_1 - cI)F_2(I + cF_2)^{-1}\|_\Delta < 1$. The Corollary then follows from Theorem 2.12.

Note that if $0 \notin \sigma(F_2)$ in $\mathscr{N}(W,W)$ then Corollary 2.12.1 requires that F_2^{-1} be incrementally exterior conic on W with center $-c$ and radius r^{-1}.

Very often the space W of interest turns out to be a Hilbert space. In that case the formulation of Corollary 2.12.1 becomes considerably more elegant, particularly if expressed in terms of sector conditions.

THEOREM 2.13

Let $F_1, F_2 \in \mathcal{N}(W,W)$ with W a Hilbert space. Then $-1 \notin \sigma(F_1 F_2)$ in $\mathcal{N}(W,W)$ if for some scalar c and $r > 0$, the operator F_1 is incrementally strictly interior conic on W with center c and radius r, $-1 \notin \sigma(cF_2)$ in $\mathcal{N}(W,W)$, and F_2 satisfies one of the following conditions:

Case 1: $|c| < r$, and F_2 is incrementally interior conic on W with

$$\text{center} \quad \frac{\bar{c}}{r^2 - |c|^2} \quad \text{and radius} \quad \frac{r}{r^2 - |c|^2}.$$

Case 2: $|c| > r$, and F_2 is incrementally exterior conic on W with

$$\text{center} \quad \frac{\bar{c}}{|c|^2 - r^2} \quad \text{and radius} \quad \frac{r}{|c|^2 - r^2}.$$

Case 3: $|c| = r$, and $I + 2cF_2$ is incrementally positive on W.

Moreover, $(I + F_1 F_2)^{-1} \in \tilde{\mathcal{B}}(W,W)$ if $(I + cF_2)^{-1} \in \tilde{\mathcal{B}}(W,W)$, and if F_1 and F_2 belong to a subalgebra of $\tilde{\mathcal{B}}(W,W)$ that contains the unit, then $(I + F_1 F_2)^{-1}$ belongs to that subalgebra if and only if $(I + cF_2)^{-1}$ does. Finally,

$$\|(I + F_1 F_2)^{-1}\|_\Delta \leqslant \frac{\|(I + cF_2)^{-1}\|_\Delta}{1 - \|F_1 - cI\|_\Delta / r} \leqslant \frac{1 + |c|/r}{1 - \|F_1 - cI\|_\Delta / r}$$

Proof: If $|c| \neq r$ then for all $x, y \in W$ the following identities hold:

$$\text{Re}\left\langle \left(F_2 - \frac{\bar{c} - r}{r^2 - |c|^2} I\right)x - \left(F_2 - \frac{\bar{c} - r}{r^2 - |c|^2} I\right)y, \right.$$

$$\left. \left(F_2 - \frac{\bar{c} + r}{r^2 - |c|^2} I\right)x - \left(F_2 - \frac{\bar{c} + r}{r^2 - |c|^2} I\right)y \right\rangle$$

$$= \|F_2 x - F_2 y\|^2 - 2 \,\text{Re}\, \frac{\bar{c}}{r^2 - |c|^2} \langle x - y, F_2 x - F_2 y\rangle$$

$$- \frac{1}{r^2 - |c|^2} \|x - y\|^2$$

$$= \frac{1}{r^2 - |c|^2}\left((r^2 - |c|^2) \|F_2 x - F_2 y\|^2\right.$$

$$\left. - 2\,\text{Re}\,\bar{c}\langle x - y, F_2 x - F_2 y\rangle - \|x - y\|^2\right),$$

$$= \frac{1}{r^2 - |c|^2}\left(r^2 \|F_2 x - F_2 y\|^2\right.$$

$$\left. - \|(I + cF_2)x - (I + cF_2)y\|^2\right).$$

It thus follows from Theorem 2.8 that both cases 1 and 2 imply that $\|(I + cF_2)x - (I + cF_2)y\| \geqslant r\,\|F_2x - F_2y\|$, which by Corollary 2.12.1 yields the conclusions of the theorem for these cases. If $|c| = r$ then $\mathrm{Re}\,\langle x - y, (I + 2cF_2)x - (I + 2cF_2)y\rangle \geqslant 0$ for all $x,\ y \in W$. Hence $\|(I + cF_2)x - (I + cF_2)y\|^2 - |c|^2\,\|F_2x - F_2y\|^2 \leqslant 0$ and $\|(I + cF_2)x - (I + cF_2)y\|^2 \geqslant r^2\,\|F_2x - F_2y\|^2$ which by Corollary 2.12.1 yields the conclusions of the theorem for case 3.

When Theorem 2.13 is expressed in terms of sector conditions, it becomes

THEOREM 2.14

Let $F_1,\ F_2 \in \mathcal{N}(W,W)$ with W a Hilbert space. Then $-1 \notin \sigma(F_1F_2)$ in $\mathcal{N}(W,W)$ if for some scalars a, b, the operator F_1 is incrementally strictly inside the sector $[a,b]$ on W, $-1 \notin \sigma(\frac{1}{2}(a + b)F_2)$ in $\mathcal{N}(W,W)$ and F_2 satisfies one of the following conditions:

Case 1: $|a + b| < |a - b|$, and F_2 is incrementally inside the sector

$$\left[-\frac{\bar{a}}{\mathrm{Re}\,\bar{a}b},\ -\frac{b}{\mathrm{Re}\,\bar{a}b}\right]$$

on W.

Case 2: $|a + b| > |a - b|$, and F_2 is incrementally outside the sector

$$\left[-\frac{b}{\mathrm{Re}\,a\bar{b}},\ -\frac{\bar{a}}{\mathrm{Re}\,\bar{a}b}\right]$$

on W.

Case 3: $|a + b| = |a - b|$, that is, $\mathrm{Re}\,a\bar{b} = 0$, and $I + (a + b)F_2$ is incrementally positive on W.

Moreover, $(I + F_2F_1)^{-1} \in \tilde{\mathcal{B}}(W,W)$ if $(I + \frac{1}{2}(a + b)F_2)^{-1} \in \tilde{\mathcal{B}}(W,W)$, and if F_1 and F_2 belong to a subalgebra of $\tilde{\mathcal{B}}(W,W)$ that contains the unit, then $(I + F_1F_2)^{-1}$ belongs to that subalgebra if and only if $(I + cF_2)^{-1}$ does.

Proof of case 1: The conditions on F_2 imply by Theorem 2.8 that F_2 is incrementally inside the cone with

$$\text{center} \quad \frac{\bar{a} + b}{2\mathrm{Re}\,a\bar{b}} \quad \text{and radius} \quad \left|\frac{\bar{a} - b}{2\mathrm{Re}\,\bar{a}b}\right|,$$

and is thus inside the cone with

$$\text{center} \quad \frac{-(\bar{a} + \bar{b})/2}{|(a + b)/2|^2 - |(a - b)/2|^2}$$

$$\text{and radius} \quad \frac{|(\bar{a} - \bar{b})/2|}{|(a - b)/2|^2 - |(a + b)/2|^2};$$

this yields the required result by Theorem 2.13.

Proof of case 2: This case is proved in a manner similar to case 1 and the details are left to the reader.

Proof of case 3: This is a direct consequence of Theorem 2.13.

The case when a and b are real is of particular interest and leads to

COROLLARY 2.14.1
Let $F_1, F_2 \in \mathcal{N}(W,W)$ with W a Hilbert space. Then $-1 \notin \sigma(F_1F_2)$ in $\mathcal{N}(W,W)$ if for some real numbers $a \leqslant b$, $b > 0$, F_1 is incrementally strictly inside the sector $[a,b]$, $-1 \notin \sigma(\frac{1}{2}(a + b)F_2)$ in $\mathcal{N}(W,W)$, and F_2 satisfies one of the following conditions:

Case 1: $a < 0$, and F_2 is incrementally inside the sector $[-1/b, -1/a]$ on W.
Case 2: $a > 0$, and F_2 is incrementally outside the sector $[-1/a, -1/b]$ on W.
Case 3: $a = 0$, and $F_2 + I/b$ is positive on W.

Moreover, $(I + F_2F_1)^{-1} \in \tilde{\mathcal{B}}(W,W)$ if $(I + \frac{1}{2}(a + b)F_2)^{-1} \in \tilde{\mathcal{B}}(W,W)$; and if F_1 and F_2 belong to a subalgebra of $\tilde{\mathcal{B}}(W,W)$ that contains the unit, then $(I + F_1F_2)^{-1}$ belongs to that subalgebra if and only if $(I + cF_2)^{-1}$ does.

The following invertibility theorem is stated in terms of positivity conditions and will play an important role in stability theory. It is an immediate consequence of the following important result that has recently appeared in the mathematical literature.

LEMMA 2.1
Let $F \in \mathcal{N}(W,W)$ be continuous on W with W a Hilbert space. If F is incrementally strictly positive on W then F is invertible on $\mathcal{N}(W,W)$ and $F^{-1} \in \tilde{\mathcal{B}}(W,W)$ and is positive on W. Moreover, if $F \in \tilde{\mathcal{B}}(W,W)$

then F^{-1} is incrementally strictly positive on W, and belongs to all subalgebras of $\tilde{\mathscr{B}}(W,W)$ which contain the unit to which F belongs.

Proof: This lemma is a consequence of the monotone mapping theorem (see Reference 11). If $F \in \tilde{\mathscr{B}}(W,W)$ then the inverse can be obtained as follows: Let $\epsilon > 0$ be such that $F - \epsilon I$ is incrementally positive on W, and let $\alpha = \epsilon/\|F\|_\Delta^2$. Then the operator $G \triangleq I - \alpha F$ is a contraction on W and $F^{-1} = (I - G)^{-1}\alpha I$. This expression then yields the sub-algebraic properties claimed for F^{-1}.

THEOREM 2.15

Let $F_1, F_2 \in \mathscr{N}(W,W)$ be continuous on W, with W a Hilbert space. Then $-1 \notin \sigma(F_1 F_2)$ in $\mathscr{N}(W,W)$ if F_1 is incrementally positive on W, F_2 is incrementally strictly positive on W, and F_2 is Lipschitz continuous or F_1 is also incrementally strictly positive. Moreover, $(I + F_1 F_2)^{-1} \in \tilde{\mathscr{B}}(W,W)$ and belongs to all subalgebras of $\tilde{\mathscr{B}}(W,W)$ which contain the unit to which F_1 and F_2 belong.

Proof: By Lemma 2.1, F_2 is invertible on W, and thus $I + F_1 F_2 = (F_2^{-1} + F_1)F_2$. Since $F_2^{-1} + F_1$ is incrementally strictly positive on W, it is invertible on W. Thus $I + F_1 F_2$ is the product of two invertible operators with inverses in $\tilde{\mathscr{B}}(W,W)$. The statement about the sub-algebras follows from the expression of the inverse given by Lemma 2.1.

Remark 1. Notice that the proof of Theorem 2.15 also yields as a side result that $F_2(I + F_1 F_2)^{-1}$ is itself incrementally strictly positive and Lipschitz continuous.

Remark 2. If F is a causal, invertible (incrementally) positive operator on W, then a simple calculation shows that F^{-1} is also (incrementally) positive. Some simple additional assumptions (see Refs. 12, 13) will then assure that F^{-1} is also a causal, invertible (incrementally) positive operator on W. These conditions lead to somewhat more general conditions of establishing the causality of inverses than the methods implied by the subalgebra considerations of Theorem 2.15.

2.8.2 *A General Invertibility Condition on W_e Involving Strongly Causal Operators*

Section 2.8.3 is concerned with invertibility conditions for causal operators on W when $S = [T_0, \infty)$. Since most of the theorems in that section *assume* a priori invertibility on W_e, it appears natural to present first a general condition for invertibility on W_e.

THEOREM 2.16

Let $S = [T_0, \infty)$ and F be a strongly causal operator on W_e. Then $\sigma(F)$ on $\mathcal{N}^+(W_e, W_e)$ consists of the zero element only.

The proof of this theorem will be given in Chapter 4 when discussing well-posedness of feedback systems.

Theorem 2.16 shows that invertibility on W_e of the identity plus a causal operator *is a very weak assumption* when the time-interval of definition S is $[T_0, \infty)$. This is a well-known fact for Volterra integral equations and differential equations.

2.8.3 *Invertibility Conditions on W for Causal Operators*

The question posed in this section is the following: Since, as pointed out in Section 2.8.2, invertibility of $I + F$ on the extended space essentially comes "for free" when F is a causal operator, can this fact somehow be exploited to obtain weaker conditions for invertibility on the nonextended space? Many operators encountered in practice are causal, and it turns out that the causality of operators can indeed be quite successfully used in obtaining such invertibility conditions. The theorems which follow have implications in stability theory, as will be pointed out in Chapter 4.

The invertibility conditions which follow are nonstandard in the mathematical literature and should be of considerable interest. The theorems are essentially restricted to causal operators, however. This does not detract from their value, since, in many fields—particularly mathematical system theory—the operators considered are more often than not causal.

The invertibility theorems which follow are completely analogous to those obtained in Section 2.8.1. However, they dispense with incremental conicity or incremental positivity in favor of conicity and positivity. The conditions thus become a great deal less restrictive.[16]

It is extremely important, when proving the theorems that follow, to keep in mind the isometrically isomorphic equivalence of $\mathcal{B}^+(W, W)$ and $\mathcal{B}^+(W_e, W_e)$ as exposed by Theorem 2.1.

---

[16] However, the resulting theorems *do not* allow one to conclude the continuity of the inverse. It is also much less obvious how to construct a recursive algorithm for the computation of the inverse. This can be done, however. For instance, in Theorem 2.17 one can use the invertibility on W_e (which is algorithmic, as will be pointed out in Chapter 4) with the invertibility on W to obtain an algorithm, a bound on the error, and a rate of convergence for the solutions. The calculations in Reference 14 may be helpful in this respect.

THEOREM 2.17

Let $F \in \mathscr{B}^+(W_e, W_e)$, and $-1 \notin \sigma(F)$ in $\mathscr{N}^+(W_e, W_e)$. If F is attenuating on W (or W_e), then $-1 \notin \sigma(F)$ in $\mathscr{B}^+(W, W)$. Moreover, $\|(I + F)^{-1}\| \leqslant (1 - \|F\|)^{-1}$.

Proof: Let $u \in W$, and consider the equation $x + Fx = u$. This equation has a unique solution $x \in W_e$ by assumption. Thus, $P_T x + P_T F x = P_T u$ for all $T \in S$, which yields

$$\|P_T x\| - \|F\| \|P_T x\| \leqslant \|P_T x\| - \|F P_T x\|$$

$$\leqslant \|P_T x\| - \|P_T F P_T x\| \leqslant \|P_T u\| \leqslant \|u\|.$$

Hence, $\|P_T x\| \leqslant \|u\|/(1 - \|F\|)$ for all $T \in S$, and $x \in W$ with $\|x\| \leqslant \|u\|/(1 - \|F\|)$. Thus $(I + F)^{-1}$ exists on W. It remains to be shown that $(I + F)^{-1}$ is causal on W. This, however, follows since $(I + F)^{-1}$ on W is the restriction of $(I + F)^{-1}$ on W_e to W. Since the latter inverse is causal by assumption, the theorem follows.

THEOREM 2.18

Let $F_1, F_2 \in \mathscr{N}^+(W_e, W_e)$, and $-1 \notin \sigma(F_1 F_2)$ in $\mathscr{N}^+(W_e, W_e)$. Then $(I + F_1 F_2)^{-1} \in \mathscr{B}^+(W, W)$ if there exists a scalar c such that $-1 \notin \sigma(cF_2)$ in $\mathscr{N}^+(W_e, W_e)$, $(I + cF_2)^{-1} \in \mathscr{B}^+(W, W)$ and $(F_1 - cI) F_2 (I + cF_2)^{-1}$ is attenuating on W. In fact,

$$\|(I + F_1 F_2)^{-1}\| \leqslant \frac{\|(I + cF_2)^{-1}\|}{1 - \|(F_1 - cI) F_2 (I + cF_2)^{-1}\|}.$$

Proof: Since $I + F_1 F_2 = [I + (F_1 - cI) F_2 (I + cF_2)^{-1}](I + cF_2)$, the operator $I + F_1 F_2$ is the product of two elements of $\mathscr{N}^+(W_e, W_e)$. Both $I + cF_2$ and $I + F_1 F_2$ are invertible in $\mathscr{N}^+(W_e, W_e)$ by assumption which thus yields that $I + (F_1 - cI) F_2 (I + cF_2)^{-1}$ is invertible in $\mathscr{N}^+(W_e, W_e)$. By Theorem 2.17, $[I + (F_1 - cI) F_2 (I + cF_2)^{-1}]^{-1}$ actually belongs to $\mathscr{B}^+(W, W)$. The estimates in the theorem are obvious.

COROLLARY 2.18.1

Let $F_1, F_2 \in \mathscr{N}^+(W_e, W_e)$, and $-1 \notin \sigma(F_1 F_2)$ in $\mathscr{N}^+(W_e, W)$. Then $(I + F_1 F_2)^{-1} \in \mathscr{B}^+(W, W)$ if for some scalar c and $r > 0$, the operator F_1 is strictly interior conic on W with center c and radius r, $-1 \notin \sigma(cF_2)$ in $\mathscr{N}^+(W_e, W_e)$, and $rF_2(I + cF_2)^{-1}$ is attenuating on W. In fact,

$$\|(I + F_1 F_2)^{-1}\| \leqslant \frac{\|(I + cF_2)^{-1}\|}{1 - \|F_1 - cI\|/r} \leqslant \frac{1 + |c|/r}{1 - \|F_1 - cI\|/r}$$

Proof: Since $(I + cF_2)(I + cF_2)^{-1} = I$ on W_e, the inverse $(I + cF_2)^{-1} = I - cF_2(I + cF_2)^{-1}$ on W_e, which shows that $(I + cF_2)^{-1} \in \mathscr{B}^+(W,W)$ and that $\|(I + cF_2)^{-1}\| \leqslant 1 + |c|/r$. Corollary 2.18.1 then follows from Theorem 2.18.

COROLLARY 2.18.2

Let $F_1, F_2 \in \mathscr{N}^+(W_e,W_e)$, and $-1 \notin \sigma(F_1F_2)$ in $\mathscr{N}^+(W_e,W_e)$. Then $(I + F_1F_2)^{-1} \in \mathscr{B}^+(W,W)$ if for some scalar c and $r > 0$, the operator F_1 is strictly interior conic on W with center c and radius r, $-1 \notin \sigma(cF_2)$ in $\mathscr{N}^+(W_e,W_e)$, and F_2 satisfies for all $x \in W$ and $T \in S$ the inequality $r \|P_T F_2 x\| \leqslant \|P_T(I + cF_2)x\|$. In fact,

$$\|(I + F_1F_2)^{-1}\| \leqslant \frac{1 + |c|/r}{1 - \|F_1 - cI\|/r}.$$

Proof: Since the conditions of the corollary imply that $rF_2(I + cF_2)^{-1}$ is attenuating on W, this theorem becomes a simple consequence of Corollary 2.18.1.

Note that if $0 \notin \sigma(F_2)$ in $\mathscr{N}^+(W_e,W_e)$ then Corollaries 2.18.1 and 2.18.2 require F_2^{-1} to be exterior conic on W_e with center $-c$ and radius r^{-1}. Note also that Corollary 2.18.2 does *not* require computation of $(I + cF)^{-1}$ or even proving that $(I + cF)^{-1} \in \mathscr{B}(W_e,W_e)$. If this has been established independently, then the condition on F_2 becomes somewhat simpler and requires that F_2 satisy, for all $x \in W$, the inequality $r \|F_2 x\| \leqslant \|(I + cF_2)x\|$.

Very often the space W under consideration is actually a Hilbert space; the formulation of the above corollaries becomes considerably more elegant if this fact is exploited, and the invertibility conditions of the previous theorems are expressed in terms of sector conditions.

THEOREM 2.19

Let $F_1, F_2 \in \mathscr{N}^+(W_e,W_e)$, with W a Hilbert space, and $-1 \notin \sigma(F_1F_2)$ in $\mathscr{N}^+(W_e,W_e)$. Then $(I + F_1F_2)^{-1} \in \mathscr{B}^+(W,W)$ if for some scalar c and $r > 0$, the operator F_1 is strictly interior conic on W with center c and radius r, $-1 \notin \sigma(cF_2)$ in $\mathscr{N}^+(W_e,W_e)$, and F_2 satisfies one of the following conditions:

Case 1: $|c| < r$, and F_2 is interior conic on W with

$$\text{center } \frac{\bar{c}}{r^2 - |c|^2} \text{ and radius } \frac{r}{r^2 - |c|^2}.$$

Case 2: $|c| > r$, and F_2 is exterior conic on W_e with

center $\dfrac{\bar{c}}{r^2 - |c|^2}$ and radius $\dfrac{r}{|c|^2 - r^2}$.

Case 3: $|c| = r$, and $I + 2cF_2$ is positive on W_e.

In fact, $\|(I + F_1F_2)^{-1}\| \leqslant \dfrac{1 + |c|/r}{1 - \|F_1 - cI\|/r}$.

Proof: Combining the ideas of the proof of Theorem 2.13 and the result of Theorem 2.18 and its corollaries leads to this result without difficulty. The details are left to the reader.

THEOREM 2.20

Let $F_1, F_2 \in \mathscr{N}^+(W_e, W_e)$, with W a Hilbert space, and $-1 \notin \sigma(F_1F_2)$ in $\mathscr{N}^+(W_e, W_e)$. Then $(I + F_1F_2)^{-1} \in \mathscr{B}^+(W,W)$ if for some scalars a, b, the operator F_1 is strictly inside the sector $[a,b]$ on W, $-1 \notin \sigma(\frac{1}{2}(a + b)F_2)$ in $\mathscr{N}^+(W_e, W_e)$, and F_2 satisfies one of the following conditions:

Case 1: $|a + b| < |a - b|$, and F_2 is inside the sector

$$\left[-\frac{\bar{a}}{\operatorname{Re} \bar{a}b}, -\frac{b}{\operatorname{Re} a\bar{b}} \right]$$

on W.

Case 2: $|a + b| > |a - b|$, and F_2 is outside the sector

$$\left[-\frac{b}{\operatorname{Re} a\bar{b}}, -\frac{\bar{a}}{\operatorname{Re} a\bar{b}} \right]$$

on W_e.

Case 3: $|a + b| = |a - b|$, i.e., $\operatorname{Re} a\bar{b} = 0$, and $I + (a + b)F_2$ is positive on W_e.

The proof of Theorem 2.20 is entirely analogous to the proof of Theorem 2.14 and is left to the reader.

The case when a and b are real is of particular interest and leads to

COROLLARY 2.20.1

Let $F_1, F_2 \in \mathscr{N}^+(W_e, W_e)$, with W a Hilbert space, and $-1 \notin \sigma(F_1F_2)$ in $\mathscr{N}^+(W_e, W_e)$. Then $(I + F_1F_2)^{-1} \in \mathscr{B}^+(W,W)$ if for some real numbers $a \leqslant b$, $b > 0$, F_1 is strictly inside the sector $[a,b]$ on W,

$-1 \notin \sigma(\frac{1}{2}(a+b)F_2)$ in $\mathcal{N}^+(W_e,W_e)$, and F_2 satisfies one of the following conditions:

Case 1: $a < 0$, and F_2 is inside the sector $[-1/b, -1/a]$ on W.
Case 2: $a > 0$, and F_2 is outside the sector $[-1/a, -1/b]$ on W_e.
Case 3: $a = 0$, and $F_2 + I/b$ is positive on W_e.

The invertibility theorems which follow involve positive operators.

THEOREM 2.21

Let $F_1, F_2 \in \mathcal{N}^+(W_e,W_e)$, with W a Hilbert space, and $-1 \notin \sigma(F_1F_2)$ in $\mathcal{N}^+(W_e,W_e)$. Then $(I + F_1F_2)^{-1} \in \mathcal{B}^+(W,W)$ if F_1 is positive on W_e and F_2 is strictly positive and bounded on W.

Proof: Let $u \in W$, and consider the equation $x + F_1F_2x = u$. This equation has a unique solution $x \in W_e$ by assumption. Thus for all $T \in S$, $P_Tx + P_TF_1F_2x = P_Tu$. Hence by strict positivity and the Cauchy inequality, there exists an $\epsilon > 0$ such that for all $T \in S$,

$$\epsilon \|P_Tx\|^2 \leqslant |\langle P_TF_2x, P_Tx + P_TF_1F_2x\rangle| = |\langle P_TF_2x, P_Tu\rangle|$$

$$\leqslant \|P_TF_2x\| \|P_Tu\| \leqslant \|F_2\| \|P_Tx\| \|P_Tu\|.$$

Consequently $\|P_Tx\| \leqslant \epsilon^{-1}\|F_2\| \|u\|$ for all $T \in S$ which shows that $x \in W$ and $\|x\| \leqslant \epsilon^{-1}\|F_2\| \|u\|$. It remains to be shown that $(I + F_1F_2)^{-1}$ is causal on W. This, however, follows since $(I + F_1F_2)^{-1}$ on W is the restriction of $(I + F_1F_2)^{-1}$ on W_e to W.

Remark: If in addition $0 \notin \sigma(F_2)$ in $\mathcal{N}^+(W_e,W_e)$ (which will happen under very weak conditions since F_2 is causal and *strictly* positive), then $F_2(I + F_1F_2)^{-1}$ turns out to be itself strictly positive and bounded on W. This fact has a rather interesting interpretation in terms of passive systems.[17] Indeed, consider F_2 and F_1 as expressing, respectively, the driving-point admittance and the driving-point impedance of a passive network. Then $F_2(I + F_1F_2)^{-1}$ is the driving-point admittance of the parallel combination shown in Figure 2.1, which is itself clearly passive if F_1 and F_2 are passive. The reader will be interested in verifying that Theorem 2.21 can be refined by requiring positivity and boundedness of F_2 (rather than of F_1) and merely positivity of F_1, or by requiring strict positivity on F_1 and F_2 but no boundedness.

[17] This interpretation is due to Zames (Ref. 10).

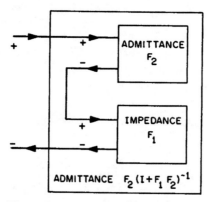

Figure 2.1   A Realization of $F_2(1 + F_1F_2)^{-1}$

2.9 Conditions for Noninvertibility of Nonlinear Operators

Section 2.8 has been concerned with establishing conditions under which nonlinear operators are invertible. The present section addresses itself to the opposite question: namely, under what conditions can it be guaranteed that a particular operator is *not* invertible? The interest in this question stems primarily from the fact that noninvertible operators lead to unstable feedback systems, a fact which will be explored in detail in Chapter 4.

A great deal of the mathematical literature concerning nonlinear operators is devoted to the question of invertibility of operators. Some such results were presented in the previous section. Quite surprisingly, virtually no research has been devoted to exploring sufficient conditions for the noninvertibility of operators. The present section constitutes such an effort and contains quite specific conditions for operators to be noninvertible.[18]

First of all, note that the invertibility theorems obtained in Section 2.8 give conditions for certain operators not to be invertible in a given subalgebra. In particular, Theorem 2.12 thus implies

[18] The estimates which go into the proofs of the noninvertibility theorems are entirely analogous to those used in the invertibility theorems of Section 2.8. The inspiration for these noninvertibility theorems stems from the relationship between noninvertibility and instability as explained in Chapter 4. The instability theorems that led the author to suspect the specific conditions are those obtained using Lyapunov theory in Reference 15. Notice also the crucial role played by Theorem 2.2 in the proof of the invertibility theorems which follow.

THEOREM 2.21

Let F_1, $F_2 \in \tilde{\mathscr{B}}^+(W,W)$. Then $-1 \in \sigma(F_1 F_2)$ in $\mathscr{N}^+(W,W)$ if there exists a scalar c such that $-1 \notin (cF_2)$ in $\tilde{\mathscr{B}}(W,W)$, $(I + cF_2)^{-1} \notin \mathscr{N}^+(W,W)$, and if $(F_1 - cI)F_2(I + cF_2)^{-1}$ is a contraction on W.

Note that Theorem 2.21 and its corollaries, which can be obtained by making the contraction condition more specific (as in Corollary 2.12.1, Theorem 2.13, Theorem 2.14, and Corollary 2.14.1) do state sufficient conditions for a particular causal operator $I + F_1 F_2$ not to be invertible in the algebra of causal operators. This is done by showing that $I + F_1 F_2$ is invertible but by ensuring, through Theorem 2.2, that this inverse is *not* causal on W. Theorem 2.21 is predicated on the possibility of showing that $I + cF_2$ is invertible on W for some scalar c, but that this inverse is not causal.

Remark 1: The preceding statement trades the problem of showing that a particular inverse $(I + F_1 F_2)^{-1}$ is not causal for a similar one, namely, showing that $(I + cF_2)^{-1}$ is not causal. In general, however, the operator $F = F_1 F_2$ is given and F_2 can be selected to convenience (modulo the condition $F = F_1 F_2$). Very often F_2 can thus be chosen to be relatively simple as compared to F, e.g., linear and time invariant, so that much more can be said a priori about the inverse $(I + cF_2)^{-1}$ than about the inverse $(I + F_1 F_2)^{-1}$.

Remark 2: When $S = [T_0, \infty)$, very weak conditions (e.g., strong causality of $F_1 F_2$) will ensure that $I + F_1 F_2$ is invertible in $\mathscr{N}^+(W_e, W_e)$, and whenever $(I + F_1 F_2)^{-1}$ exists in $\mathscr{N}(W,W)$, it will thus necessarily be causal (as the restriction of this inverse on W_e to W). Theorem 2.21 will then, of course, never succeed in showing that $I + F_1 F_2$ is not invertible in $\mathscr{N}^+(W,W)$. A method, believed to be innovative, of treating this case is outlined in the remainder of this section.

Definition: Let $S = [T_0, \infty)$, W (defined over S), and $F \in \mathscr{N}^+(W_e, W_e)$ be given. Let W' consist of B-valued functions defined on $S' = (-\infty, +\infty)$ and $F' \in \mathscr{N}^+(W'_e, W'_e)$. Then S', W', and F' are said to be a *backwards extension*[19] of S, W, and F if they satisfy the following hypotheses:

1. The space W' satisfies the assumptions W.1–W.4 enumerated in Section 2.3.

---

[19] Compare this with the definition of the restriction of an operator to a smaller time-interval of definition, as explained in Section 2.7. In fact, F is a restriction of F' if and only if F' is a backwards extension of F.

2. If $x \in W$, then the B-valued function

$$x'(t) = \begin{cases} x(t) & \text{for} & t \in S \\ 0 & \text{for} & t \in (S' - S) \end{cases}$$

belongs to W'.
3. If $x' \in W'$, then the B-valued function x with $x(t) = x'(t)$ for $t \in S$ belongs to W.
4. The operator F' is an element of $\mathcal{N}^+(W'_e, W'_e)$.
5. Given any $x' \in W'$ with $x'(t) = 0$ for $t < T_0$ and $x \in W$ with $x(t) = x'(t)$ for $t \in S$, then $F'x'(t) = Fx(t)$ for $t \in S$.

The following lemma plays a crucial role in the results that follow and is believed to be of some interest in its own right. It essentially states that if a Lipschitz-continuous operator is invertible on one half-line, then it is invertible on any half-line.

LEMMA 2.2

Let t_1, $t_2 \in S$ be given, $F \in \tilde{\mathcal{B}}^+(W_e, W_e)$, and let F be invertible in $\mathcal{N}^+(W_e, W_e)$. Then F has an inverse on $W_{t_1} \triangleq \{x \in W \mid P_{t_1}x = 0\}$ if and only if F has an inverse on $W_{t_2} \triangleq \{x \in W \mid P_{t_2}x = 0\}$.

Proof: Note that F has a *causal* inverse on $W_{t_1 e}$ and $W_{t_2 e}$ since it has a causal inverse on W_e by assumption. Let $t_1 \leqslant t_2$. If F has an inverse on W_{t_1}, then the restriction of this inverse to the subspace W_{t_2} of W_{t_1} clearly qualifies as the inverse on W_{t_2} since by causality this inverse maps W_{t_2} into itself. Conversely, assume that F is invertible on W_{t_2} and let $u \in W_{t_1}$ be given. Let $e = F^{-1}u \in W_{t_1 e}$ be decomposed as $e = P_{t_2}e + (I - P_{t_2})e$. Thus $P_{t_2}e \in W$. Since F is Lipschitz continuous, it follows that $\|P_T(Fe - F(I - P_{t_2})e)\| \leqslant \|F\|_\Delta \|P_{t_2}e\|$ for all $T \in S, T \geqslant t_2$, which shows that $v \triangleq Fe - F(I - P_{t_2})e \in W$. Hence $F(I - P_{t_2})e = Fe - v = u - v \in W$. Since $u - v \in W_{t_2}$ and F is invertible on W_{t_2}, this thus yields $(I - P_{t_2})e = F^{-1}(u - v) \in W_{t_2}$. Thus $e = P_{t_2}e + (I - P_{t_2})e \in W$, which ends the proof of the lemma.

THEOREM 2.22

Let $S = [T_0, \infty)$, $F \in \tilde{\mathcal{B}}^+(W_e, W_e)$, and F be invertible in $\mathcal{N}^+(W_e, W_e)$. Let W' and $F' \in \tilde{\mathcal{B}}^+(W'_e, W'_e)$ be a backwards extension of W and F from $S = [T_0, \infty)$ to $S' = (-\infty, +\infty)$, and assume that for all $T \in S'$ the operator F' has a causal inverse on $W'_{Te} \triangleq \{x \in W'_e \mid P_Tx = 0\}$. Then $F^{-1} \notin \mathcal{N}(W, W)$ if F' has a noncausal inverse on W'_{T_0}.

Proof: Since F' has a noncausal inverse on W', there exists at least one $T \in S'$ such that F' is not invertible on $W'_T \triangleq \{x \in W' \mid P_T x = 0\}$. This then implies by Lemma 2.2 that F' is not invertible on $W'_{T_0} \triangleq \{x \in W' \mid P_{T_0} x = 0\}$. Since W' and F' are backward extensions of W and F, the spaces W'_{T_0} and W are isomorphic and F' is equivalent to F in the sense that F' and F map equivalent elements $x' \in W'_{T_0}$ and $x \in W$ into equivalent elements $F'x' \in W'_{T_0}$ and $Fx \in W$. Thus F is not one-to-one and onto W if F' is not one-to-one and onto W'. Hence F is not invertible on W as claimed.

Theorems 2.21 and 2.22 can then be combined to give a rather concrete noninvertibility theorem. More specifically, they yield

THEOREM 2.23

Let $S = [T_0, \infty)$, $F \in \tilde{\mathscr{B}}^+(W_e, W_e)$, and let F be invertible in $\mathscr{N}^+(W_e, W_e)$. Let W' and $F' \in \tilde{\mathscr{B}}^+(W'_e, W'_e)$ be a backwards extension of W and F from $S = [T_0, \infty)$ to $S' = (-\infty, +\infty)$, and assume that for all $T \in S'$, F' has a causal inverse on $W'_{T_e} \triangleq \{x \in W'_e \mid P_T x = 0\}$. Let $F'_1, F'_2 \in \tilde{\mathscr{B}}^+(W', W')$ and $F' = I + F'_1 F'_2$. Then $F^{-1} \notin \mathscr{N}(W, W)$ if there exists a scalar c such that $-1 \notin \sigma(cF'_2)$ in $\tilde{\mathscr{B}}(W', W')$, $(I + cF'_2)^{-1} \notin \mathscr{N}^+(W', W')$, and if $(F'_1 - cI)F'_2(I + cF'_2)^{-1}$ is a contraction on W'.

It is again possible to make the contraction condition of Theorem 2.23 more specific and obtain the analogues of Corollary 2.12.1, Theorem 2.13, Theorem 2.14, and Corollary 2.14.1. These details are left to the reader. Notice, however, that only the case of exterior conicity or outside the sector conditions on F_2 will then lead to non-invertibility conditions.

It should be emphasized at this point that it is in general *not* difficult to obtain suitable backwards extensions of particular operators. For time-invariant operators the obvious choice to consider is the (unique) backward extension which itself is time invariant. Examples illustrating the use of Theorem 2.23 will be given in connection with instability theorems in Chapter 5.

References

1. Windeknecht, T. G., "Mathematical System Theory—Causality," *Math. System Theory*, Vol. 1, No. 4, pp. 279–288, 1967.
2. Saeks, R., *Causality in Hilbert Space*, Electrical Engineering Memorandum EE-6814-c, University of Notre Dame, Notre Dame, 1969.

3. Porter, W. A., and Zahm, C. L., *Basic Concepts in System Theory*, SEL Technical Report 44, University of Michigan, Ann Arbor, 1969.
4. Hille, E., and Phillips, R. S., *Functional Analysis and Semi-Groups* (second edition), American Mathematical Society, Providence, 1957.
5. Falb, P. L., and Freedman, M. I., "A Generalized Transform Theory for Causal Operators," *SIAM J. on Control*, Vol. 7, No. 3, pp. 452–471, 1969.
6. Falb, P. L., Freedman, M. I., and Zames, G., "A Hilbert Space Stability Theory Over Locally Compact Abelian Groups," *SIAM J. on Control*, Vol. 7, No. 3, pp. 479–495, 1969.
7. Lumer, G., and Phillips, R. S., "Dissipative Operators in a Banach Space," *Pacific J. Math.*, Vol. 11, pp. 679–698, 1961.
8. Saaty, T. L., *Modern Nonlinear Equations*, McGraw-Hill, New York, 1967.
9. Shinbroth, M., "Fixed Point Theorems," *Scientific American*, January 1966.
10. Zames, G., "On the Input-Output Stability of Time-Varying Nonlinear Feedback Systems. Part I: Conditions Derived Using Concepts of Loop Gain, Conicity, and Positivity. Part II: Conditions Involving Circles in the Frequency Plane and Sector Nonlinearities," *IEEE Trans. on Automatic Control*, Vol. AC-11, pp. 228–238 and 465–476, 1966.
11. Minty, G. J., "Monotone (Nonlinear) Operators in Hilbert Space," *Duke Math. J.*, Vol. 29, pp. 341–346, 1962.
12. Youla, D. C., Castriota, L. J., and Carlin, H. J., "Bounded Real Scattering Matrices and the Foundations of Linear Passive Network Theory, *IRE Trans. on Circuit Theory*, Vol. CT-6, pp. 102–124, 1959.
13. Sandberg, I. W., "Conditions for the Causality of Nonlinear Operators Defined on a Linear Space," *Quart. of Appl. Math.*, Vol. 23, No. 1, pp. 87–91, 1965.
14. Chu, S. C., and Diaz, J. B., "On in the Large Application of the Contraction Principle," in *Differential Equations and Dynamical Systems* (edited by LaSalle and Hale), pp. 235–238, Academic Press, New York, 1967.
15. Brockett, R. W., and Lee, H. B., "Frequency Domain Instability Criteria for Time-Varying and Nonlinear Systems," *IEEE Proceedings*, Vol. 55, pp. 604–619, 1967.

3 Positive Operators

3.1 Introduction

This chapter is devoted to the study of positive operators. An operator will be called "positive" if the real part of the inner product of any element and its image under the operation is nonnegative. For example, a linear transformation from a finite-dimensional real vector space into itself defines a positive operator if and only if the matrix associated with this linear transformation plus the transpose of the matrix is nonnegative definite. The Sylvester test yields a simple necessary and sufficient condition for a finite-dimensional linear transformation to define a positive operator. For nonlinear transformations or operators defined on infinite-dimensional spaces, the situation is quite different and this is where the techniques and results developed in this chapter are useful.

Why are positive operators important? There are several areas both in engineering and in applied mathematics where positive operators play a central role. Here are some examples:

1. Many techniques—e.g., in the theory of optimal control, in prediction theory, and in stability theory—require at some point in the analysis *that a certain function or functional be positive definite;* instances of this are the second variations in optimization theory and Lyapunov functions and their derivatives in stability theory. This procedure often reduces to one of verifying that a certain appropriately chosen operator is positive. In this context, it suffices to recall how often the positive definiteness of certain matrices is invoked.

45

2. Another area of research where positive operators play an essential role is *network synthesis*. Recall that a ratio of polynomials in the complex variable *s* is the driving-point impedance of a two-terminal network that can be realized using a finite number of positive, linear, and constant resistors, inductors, and capacitors, if and only if this ratio of polynomials is a positive real function of *s*. This result thus identifies the input-output relation of these passive networks with a class of positive operators.[1] It leaves no doubt that positive operators will also play an essential role in the synthesis of nonlinear and time-varying networks using certain passive devices.

3. An important application of positive operators is in establishing the *stability of feedback systems*. Roughly speaking, stability is the property of systems whereby small inputs or initial conditions produce small responses. The technique for generating stability criteria for feedback systems from knowledge of positive operators will be examined in detail in Chapter 4, but the basic idea is simple and states that the interconnection of passive systems (positive operators) yields a stable system.

4. The so-called *frequency-power formulas* have been applied to the design of parametric amplifiers. They are formulas which constrain weighted sums of real and reactive powers entering a device at various frequencies to be either zero, positive, or negative. The device could be, for instance, a nonlinear resistor, inductor, or capacitor. This work was initiated by Manley and Rowe, who analyzed the power flow at various frequencies in a nonlinear capacitor and discovered the now-famous Manley-Rowe frequency-power formulas. Manley and Rowe's work has been extended in several directions and the resulting formulas have found application in the design of frequency converters. Frequency-power formulas establish fundamental limits on the efficiency of such devices. These formulas have been applied in energy conversion using parametric devices, in studies of hydrodynamic and magneto-hydrodynamic stability, and in many other areas. In trying to bring certain methods and results in these areas into harmony, it became apparent that these frequency-power formulas are essentially particular classes of positive operators and can, mathematically at least, be most easily understood as such.

5. Another important application of positive operators is the determination of *bounds on the optimal performance* of nonlinear time-varying systems. One of the crucial problems in optimal control

[1] This relationship between positivity of operators and passivity of systems has motivated many authors to call positive operators *passive* or *dissipative*.

theory is, paradoxically, the design of suboptimal systems. In fact, because of computational feasibility and more convenient implementation, it is in many cases necessary to resort to suboptimal systems. Little or no attention has been paid to the problem of a priori predicting how far a suboptimal system is from being optimal. It can be shown[2] that the requirement that a given system has a better performance than another with respect to some performance criterion can in many important cases be reduced to requiring that a certain suitably chosen operator be positive. This then allows one to estimate a priori bounds on the performance of certain systems and to design feasible suboptimal controls. In this respect it is also worthwhile to mention that optimal control provides, conceptually at least, a way of verifying the positivity of an operator F. Indeed if $\inf_x \operatorname{Re} \langle x, Fx \rangle \geqslant 0$, then the operator is clearly positive.

6. From the purely mathematical point of view, positive operators are important because of their implications about the *invertibility of nonlinear operators*. These aspects have been explored in Chapter 2.

The first class of operators which are examined in this chapter for positivity are the convolution operators and the memoryless operators in which the output is an instantaneous function of the input. These positive operators lead to the well-known Manley-Rowe equations and play an important role in stability theory since they are closely connected with the Popov criterion and the circle criterion for the stability of nonlinear and time-varying feedback systems.

Next, attention is focused on the question what class of linear convolution operators can be composed with a positive periodically time-varying linear multiplicative gain and still yield a positive operator. The answer to this question turns out to require that this convolution operator itself be positive and that the kernel of the convolution should consist of a string of impulses occurring at multiples of the period of the time-varying gain. It is also shown that this result is the best of its type.

In Section 3.2, an answer to the following question is sought: What is the most general linear operator that when composed with a monotone nondecreasing (or an odd-monotone nondecreasing) nonlinearity yields a positive operator? This problem has received a great deal of attention in the past, in connection with both the frequency-power

---

[2] The topic of a priori bounds of suboptimal controllers is treated in Reference 1. This work brings out the relevance of positivity conditions to this important area of research.

formulas and the stability of feedback loops with a monotone non-linearity in the feedback loop. The resulting class of positive operators is closely related to certain classes of matrices, i.e., the dominant matrices, which are important in network synthesis. The mathematical reason for this connection, however, remains vague and deserves further investigation. As an intermediate step in deriving this class of positive operators a considerable generalization of a classical inequality due to Hardy, Littlewood and Polya on the rearrangement of sequences is obtained. It is felt that the extension of this rearrangement inequality is of intrinsic importance in itself and has applications in other areas of system theory.

Section 3.8 is devoted to the problem of adjoining to a positive operator a causal positive operator. It is shown that it is possible to adjoin a causal positive operator with an arbitrary positive operator provided the operator admits a suitable factorization. Whether a particular operator satisfies this condition appears to have no general answer, and the problem is one of considerable interest and importance. Similar factorizations have received a great deal of attention in the past, particularly in the classical prediction-theory literature. In this section a general factorization theorem is presented which is felt to be quite general and of intrinsic importance. The result, which is based on contraction arguments, unfortunately does not offer a necessary condition and is rather conservative in some particular cases.

Recall the definition of a positive operator: Let X be an inner product space and let F be a mapping from Do $(F) \subset X$ into X. Then F is said to be (*incrementally*) *positive* on Do (F) if for all $x \in$ Do (F) $(x, y \in$ Do $(F))$, Re $\langle x, Fx \rangle \geqslant 0$ (Re $\langle x - y, Fx - Fy \rangle \geqslant 0$). It is said to be (*incrementally*) *strictly positive* on Do (F) if for some $\epsilon > 0$, $F - \epsilon I$ is (incrementally) positive on Do (F).

Some elementary properties of positive operators are:

1. The (incrementally) positive operators form a cone; i.e., if F_1 and F_2 are positive on Do (F_1) and Do (F_2), respectively, then $F_1 + F_2$ is positive on Do $(F_1) \cap$ Do (F_2), and if α is a nonnegative real number, then αF_1 is positive on Do (F_1).
2. The (incrementally) positive operators are closed under inversion; i.e., if F is positive on Do (F), then F^{-1} is positive on Do (F^{-1}).
3. The (incrementally) positive linear operators are closed under the adjoint operation; i.e., if L is a linear operator from Do (L) into X and if L^* denotes the adjoint, then L is positive on Do (L) if and only if L^* is positive on Do (L^*).

4. If F is positive on its domain and L is linear then L^*FL is positive on its domain.

3.2 Positive Convolution Operators and Positive Memoryless Operators

The space under consideration in this section is $L_2^H(S)$ (denoted simply by L_2 when no confusion can occur) with H a given Hilbert space and $S = (-\infty, +\infty)$. The case $H = R$ is of particular interest and will be the subject of a number of corollaries. As mentioned in Chapter 1, the space L_2 is a Hilbert space with $\langle x, y \rangle_{L_2} \triangleq \int_{-\infty}^{+\infty} \langle x(t), y(t) \rangle_H \, dt$. Recall that the limit-in-the-mean transform of x, defined by $\hat{X}(j\omega) \triangleq \text{l.i.m.} \int_{-\infty}^{+\infty} x(t) e^{-j\omega t} \, dt$, is then, for any $x \in L_2$, a well-defined, H-valued function of ω, $\omega \in R$, which itself belongs to $L_2^H(-\infty, +\infty)$. Moreover, the transform of $\hat{X} = x/2\pi$ and Parseval's equality states that $\langle x, y \rangle_{L_2} = 2\pi \langle \hat{X}, \hat{Y} \rangle_{L_2}$ for all $x, y \in L_2$.

Definitions: Let $\mathscr{G}$ denote the class of operators from L_2 into itself defined as follows: associated with each element $G \in \mathscr{G}$ is a class of operators $G(j\omega) \in \mathscr{L}(H,H)$, parametrized by ω, $\omega \in R$, with $G(j\omega) \in L_\infty$, and which maps the element $x \in L_2$ into the element of L_2 with limit-in-the-mean transform $G(j\omega)\hat{X}(j\omega)$.

It is simple to verify that the operator G is indeed well defined; i.e., that it maps L_2 into itself. An important subclass of operators of this type are the *convolution operators,* which are defined as follows: let $g(t)$, $t \in R$, and g_k, $k \in I$, be elements of $\mathscr{L}(H,H)$ with $(\|g(t)\|, \{\|g_k\|\}) \in L_1 \times l_1$, and let $\{t_k\}$ be a mapping from I into R. Let $y(t) = (Gx)(t)$ be formally defined by

$$y(t) \triangleq \sum_{k=-\infty}^{+\infty} g_k x(t - t_k) + \int_{-\infty}^{+\infty} g(t - \tau) x(\tau) \, d\tau.$$

Lemma 3.1

The convolution operator G formally defined above maps L_2 into itself. Moreover $G \in \mathscr{G}$ and the function $G(j\omega)$ associated with G is given by

$$G(j\omega) \triangleq \sum_{k=-\infty}^{+\infty} g_k e^{-j\omega t} + \int_{-\infty}^{+\infty} g(t) e^{-j\omega t} \, dt.$$

This is a standard result from Fourier transform theory (see Ref. 2, p. 90).[3] Note that G is causal on L_2 if and only if $g(t) = 0$ for almost all $t < 0$ and $t_k \geqslant 0$ for all $k \in I$.

[3] Lemma 3.1 remains valid if $g(t) \in L_2$ and $\{g_k\} \in l_2$ both have limit-in-the-mean transforms in L_∞.

THEOREM 3.1

Every element $G \in \mathcal{G}$ defines a time-invariant bounded linear operator on L_2, $\|G\| = \|G(j\omega)\|_{L_\infty}$, and G is a positive operator on L_2 if and only if for almost all $\omega \in R$, $G(j\omega)$ is (pointwise) positive on H. Moreover, $G^* \in \mathcal{G}$, and has the function $G^*(j\omega)$ associated with it.

Proof: The theorem is obvious with the possible exception of the positivity condition and the time invariance. Positivity follows from Parseval's equality since

$$\mathrm{Re} \, \langle x(t), Gx(t) \rangle = \frac{1}{2\pi} \int_{-\infty}^{+\infty} \mathrm{Re} \, \langle \hat{X}(j\omega), G(j\omega)\hat{X}(j\omega) \rangle_H \, d\omega.$$

Time invariance is proved as follows: let $\tau \in R$, $x \in L_2$, and $(T_\tau x)(t) \triangleq x(t + \tau)$. Then $\widehat{GT_\tau x} = G(j\omega)\hat{X}(j\omega)e^{j\omega\tau}$ which clearly equals $\widehat{T_\tau Gx}$.

COROLLARY 3.1.1

Let $H = R$ and $G(j\omega) = \bar{G}(-j\omega)$. Then G is positive on L_2 if and only if $\mathrm{Re} \, G(j\omega) \geqslant 0$ for almost all $\omega \geqslant 0$.

Definition: Let $\mathcal{K}$ denote the class of operators[4] from L_2 into itself defined as follows. Associated with each element $K \in \mathcal{K}$ is a class of operators $K(t) \in \mathcal{L}(H,H)$, parametrized by t, $t \in R$, with $K(t) \in L_\infty$, and which maps the element $x \in L_2$ into the element of L_2 defined by $(Kx)(t) \triangleq K(t)x(t)$. From this definition there follows

THEOREM 3.2

Every element $K \in \mathcal{K}$ defines a bounded memoryless linear operator on L_2, $\|K\| = \|K(t)\|_{L_\infty}$, and K is a positive operator on L_2 if and only if $K(t)$ is (pointwise) positive on H for almost all $t \in R$. Moreover, $K^* \in \mathcal{K}$ and has the function $K^*(t)$ associated with it.

COROLLARY 3.2.1

If $H = R$, then every element $K \in \mathcal{K}$ is self-adjoint and K is positive on L_2 if and only if $K(t) \geqslant 0$ for almost all $t \in R$.

---

[4] The classes $\mathcal{G}$ and $\mathcal{K}$ are dual in the sense that both constitute pointwise multiplications; the former in the frequency domain and the latter in the time domain. This duality permeates the recent work on stability from an input-output point of view.

Definition: Let $\mathscr{F}$ denote the class of operators from L_2 into itself defined as follows. Associated with each element $F \in \mathscr{F}$ is a class of operators $F(\cdot,t)$, parametrized by t, $t \in R$, each mapping H into itself, such that there exists a constant $M < \infty$ such that $\|F(x,t)\| \leqslant M \|x\|$ for all $x \in H$ and $t \in R$, and which maps the element $x \in L_2$ into the element of L_2 defined by $(Fx)(t) \triangleq F(x(t),t)$. There follows

THEOREM 3.3

Every element $F \in \mathscr{F}$ defines a bounded memoryless operator on L_2, $\|F\| = M'$, with $M' = \inf M$ over all M such that for all $x \in H$ and almost all $t \in S$, $\|F(x,t)\| < M \|x\|$; i.e., $\|F\| = (\|F(\cdot,t)\|)_{L_\infty}$, and F is (incrementally) positive on L_2 if and only if $F(\cdot,t)$ is (incrementally) positive on H for almost all $t \in R$. The operator F is Lipschitz continuous on L_2 if $F(\cdot,t)$ is Lipschitz continuous for almost all $t \in S$, uniformly in t, and $\|F\|_\Delta = (\|F(\cdot,t)\|_\Delta)_{L_\infty}$.

3.3 Memoryless Time-Invariant Nonlinear Operators

An important particular class of operators in $\mathscr{F}$ are those for which $F(\cdot,t)$ is constant for almost all $t \in R$. The operator $F \in \mathscr{F}$ associated with $F(\cdot,t)$ then becomes time invariant. Attention is now focused on this case, with $H = R$. It turns out that a number of interesting relations can be obtained. These results play in fact an important role in the design of frequency converters where they are known under the name of the Manley-Rowe equations and in the stability of nonlinear feedback systems where they lead to the Popov criterion.

The operators under investigation are thus the time-invariant operators in $\mathscr{F}$ with $H = R$ and $S = (-\infty, +\infty)$. Let this class be denoted by $\tilde{\mathscr{F}}$. Each element of $\tilde{\mathscr{F}}$ has thus associated with it a map $f(\cdot)$ from the real line into itself, satisfying, for some constant $M < \infty$ and all $\sigma \in R$, the inequality $|f(\sigma)| \leqslant M |\sigma|$, and $F \in \tilde{\mathscr{F}}$ maps L_2 into itself according to $(Fx)(t) = f(x(t))$.

Definition: A function x from R into itself is said to be *absolutely continuous* if for any integer N and any sequence $\{t_k\}$, $k = 1, 2, \ldots, N$,

$$\sum_{k=1}^{N-1} |x(t_k) - x(t_{k+1})| \to 0 \quad \text{whenever} \quad \sum_{k=1}^{N-1} |t_k - t_{k+1}| \to 0.$$

A classic result in analysis states that a function is absolutely continuous if and only if $x(t) = x(a) + \int_a^t r(t)\, dt$ for some function $r(t) \in L_1(a,b)$. The function $x(t)$ is then differentiable almost everywhere, and $r(t) = \dot{x}(t)$ for almost all t. Let $S_2^1(a,b)$ be the subspace of $L_2(a,b)$ formed by the functions on $[a,b]$ that are absolutely continuous and that belong, together with their derivatives, to $L_2(a,b)$. Let S_2^1 denote $S_2^1(-\infty,+\infty)$. The space S_2^1 is an inner product space with the inner product as in L_2. It is, however, not closed in the L_2-topology.

LEMMA 3.2

If $x \in S_2^1$ then $\lim_{t\to\pm\infty} x(t) = 0$.

Proof: Write $\int_{-T_1}^{T_2} x(t)\dot{x}(t)\,dt = \frac{1}{2}[x^2(T_2) - x^2(-T_1)]$; it follows that the limits exist, since the limit of the integral for T_1 or $T_2 \to \infty$ exists by the Schwartz inequality. Since $\lim_{t\to\pm\infty} x(t)$ thus exist and since $x(t) \in L_2$, the limits must indeed be zero.

THEOREM 3.4

Assume that $F \in \tilde{\mathscr{F}}$ and that the function f that defines F is Lipschitz continuous on R. Then $\langle x, d(Fx)/dt\rangle_{L_2} = 0$ for all $x \in S_2^1$.

Proof: Let $y(t) = Fx(t)$ and let K be a Lipschitz constant for f. Since f is Lipschitz continuous, $y(t)$ is absolutely continuous whenever $x(t)$ is. It is simple to show that $|\dot{y}(t)| \leq K|\dot{x}(t)|$ whenever both exist (and thus almost everywhere). The inner product in the theorem statement is thus well defined since $y \in S_2^1$. Integration by parts yields

$$\int_{-\infty}^{+\infty} x(t)\frac{d}{dt}y(t)\,dt = -\int_{-\infty}^{+\infty} f(x(t))\frac{d}{dt}x(t)\,dt$$

$$= -\lim_{T\to\infty}\int_{x(-T)}^{x(T)} f(\sigma)\,d\sigma$$

$$= 0$$

The last equality follows from Lemma 3.2.

Remark: If $x \in S_2^1$, then the limit-in-the-mean transform of $\dot{x}$ exists and equals $j\omega\hat{X}(j\omega)$. Thus Theorem 3.4 merely states that for all $x \in L_2$ and $y = Fx$

$$\int_{-\infty}^{+\infty} j\omega\,\hat{X}(-j\omega)\hat{Y}(j\omega)\,d\omega = 0,$$

which is precisely the Manley-Rowe power-frequency formula[5] for elements of L_2.

[5] The Manley-Rowe equation thus merely states that the vectors $\dot{x}$ and Fx are orthogonal in L_2.

Combining Theorems 3.3 and 3.4 yields

Theorem 3.5

Let $F \in \tilde{\mathscr{F}}$, and assume that the function f which determines F is Lipschitz continuous on R. Let $\alpha \in R$ and $G \in \mathscr{G}$ be defined by $G(j\omega) = 1/(1 + \alpha j\omega)$. Then FG is a nonnegative operator on L_2 if and only if for all $\sigma \in R$, $\sigma f(\sigma) \geqslant 0$.

Proof: The theorem is a particular case of Theorem 3.3 if $\alpha = 0$. Let therefore $\alpha \neq 0$. Since the operator G corresponds to a convolution, Gx is absolutely continuous for all $x \in L_2$. Moreover, since $j\omega/(1 + \alpha j\omega) \in L_\infty$ for $\alpha \neq 0$, $Gx \in S_2^1$ for all $x \in L_2$. Thus

$$\langle x, FGx \rangle = \left\langle \left(1 + \alpha \frac{d}{dt}\right) Gx, FGx \right\rangle$$

$$= \langle Gx, FGx \rangle + \alpha \left\langle \frac{d}{dt} Gx, FGx \right\rangle \geqslant \alpha \left\langle \frac{d}{dt} Gx, FGx \right\rangle.$$

This last inner product equals zero by Theorem 3.4. The converse part of the theorem is an immediate consequence of Theorems 3.3 and 3.4.

3.4 Periodic Gain

The result in this section concerns a positive operator formed by the interconnection of a periodically time-varying gain and a linear time-invariant convolution operator. The proof is very simple, and this positive operator leads to a rather elegant frequency-domain stability criterion that will be discussed in Chapter 6.

Definitions: Let T be a positive real number, and let $\mathscr{K}_T$ denote the subclass of $\mathscr{K}$ determined by functions $k(t)$ which satisfy $k(t + T) = k(t)$ for almost all t. Let $\mathscr{G}_T$ denote the subclass of elements of $\mathscr{G}$ determined by functions $G(j\omega)$ which satisfy $G(j(\omega + 2\pi T^{-1})) = G(j\omega)$ for almost all ω.

It is again assumed that $H = R$. Each element of $\mathscr{K}_T$ thus has associated with it a real-valued function $k(t)$ defined on $(-\infty, +\infty)$, with $k(t + T) = k(t)$ for almost all $t \in R$, $k(t) \in L_\infty$, and which maps L_2 into itself according to $(Kx)(t) = k(t)x(t)$. An operator $K \in \mathscr{K}_T$ thus corresponds to a *periodic gain* with period T. Each element of $\mathscr{G}_T$ has associated with it a complex-valued function $G(j\omega) = \bar{G}(-j\omega)$ defined for $\omega \in R$, with $G(j(\omega + 2\pi T^{-1})) = G(j\omega)$ for almost all ω, $G(j\omega) \in L_\infty$, and which maps the element $x \in L_2$ into the element with limit-in-the-mean transform $G(j\omega)\hat{X}(j\omega)$. An important subclass of operators in $\mathscr{G}_T$ are those defined by the convolution with kernel $\sum_{n \in I} g_n \, \delta(t - nT)$, where $\{g_n\} \in l_1$, and δ denotes the Dirac delta function.

LEMMA 3.3

Let $K \in \mathscr{K}_T$ and $G \in \mathscr{G}_T$. Then K and G commute on L_2, i.e., $KGx = GKx$ for all $x \in L_2$.

Proof:[6] Since both K and G are bounded linear operators from L_2 into itself, KG and GK are. By continuity of bounded linear operators, it thus suffices to prove the lemma for a dense set in L_2. Define the sequence $\{g_n\}$, $n \in I$, by

$$g_n \triangleq \frac{T}{2\pi} \int_0^{2\pi/T} G(j\omega)^{-jn\omega T} d\omega.$$

It follows from the theory of Fourier series that $\{g_n\} \in l_2$ and that

$$G(j\omega) = \text{l.i.m.} \sum_{n=-\infty}^{+\infty} g_n e^{jn\omega T} = \lim_{N \to \infty} \sum_{n=-N}^{+N} g_n e^{jn\omega T}.$$

Let v be any element of $L_2 \cap L_1$. Then

$$w_N(t) = \sum_{n=-N}^{N} g_n v(t - nT) \in L_2 \cap L_1.$$

Let $\hat{V}$ and $\hat{W}_N$ denote the limit-in-the-mean transforms of v and w_N respectively. Then

$$\hat{W}_N(j\omega) = \sum_{n=-N}^{N} g_n e^{jn\omega T} \hat{V}(j\omega).$$

Thus

$$\|\hat{W}_N(j\omega) - G(j\omega)\hat{V}(j\omega)\|_{L_2} = \left\|\left(G(j\omega) - \sum_{n=-N}^{N} g_n e^{jn\omega T}\right)\hat{V}(j\omega)\right\|_{L_2}$$

Since $\hat{V}(j\omega) \in L_\infty$ and $\|\hat{V}(j\omega)\|_{L_\infty} \leqslant \|v\|_{L_1}$, it follows from Hölder's inequality that

$$\|W_N(j\omega) - G(j\omega)V(j\omega)\|_{L_2} \leqslant \|v\|_{L_1} \|G(j\omega) - \sum_{n=-N}^{N} g_n e^{jn\omega T}\|_{L_2}.$$

[6] The conclusion of Lemma 3.3 is immediate if one is satisfied with the following formal argument: since $Gx(t) = \sum_{n=-\infty}^{+\infty} g_n x(t - nT)$ and $k(t) = k(t - nT)$, then

$$KGx(t) = k(t) \sum_{n=-\infty}^{+\infty} g_n x(t - nT) = \sum_{n=-\infty}^{+\infty} g_n k(t - nT) x(t - nT) = GKx(t).$$

Since $\sum_{n=-N}^{N} g_k e^{jn\omega T}$ approaches $G(j\omega)$ in the L_2 sense, this shows that w_N approaches in the L_2 sense the function whose limit-in-the-mean transform equals $G(j\omega)\hat{V}(j\omega)$. Hence

$$w(t) = \text{l.i.m.} \sum_{n=-\infty}^{+\infty} g_n v(t - nT)$$

exists, belongs to L_2, and has $G(j\omega)\hat{V}(j\omega)$ as its limit-in-the-mean transform. This holds for all $v \in L_2 \cap L_1$. The lemma will now be proved for all $x \in L_2 \cap L_1$. Since then $Kx \in L_2 \cap L_1$, the above analysis applies to both x and Kx. However, $k(t)g_n x(t - nT) = g_n k(t - nT)x(t - nT)$ for all $n \in I$, and thus

$$k(t) \sum_{n=-N}^{N} g_n x(t - nT) = \sum_{n=-N}^{N} g_n k(t - nT)x(t - nT),$$

which, after taking the limit-in-the-mean of both sides and observing that $k \in L_\infty$, yields $KGx = GKx$ for all $x \in L_2 \cap L_1$. Since $L_2 \cap L_1$ is dense in L_2, the lemma follows.

Definitions: An operator F from L_2 into itself is said to possess a *square root*, denoted by $F^{1/2}$, if there exists an operator, $F^{1/2}$, from L_2 into itself such that $F = F^{1/2}F^{1/2}$.

LEMMA 3.4

Let $K \in \mathcal{K}$ be determined by $k(t)$, and assume that for almost all $t \in R$, $k(t) \geqslant 0$. Then $K^{1/2}$ exists. Moreover, $K^{1/2} \in \mathcal{K}$ and $K^{1/2} \in \mathcal{K}_T$ if $K \in \mathcal{K}_T$.

Proof: The element of $\mathcal{K}$ determined by $k(t)^{1/2}$ possesses all the required properties.

THEOREM 3.6

Let $K \in \mathcal{K}_T$ and let $G \in \mathcal{G}_T$. Then KG and GK are (strictly) positive operators on L_2 if for almost all $t \in R$ and $\omega \geqslant 0$, $k(t) \geqslant 0$ and $\text{Re } G(j\omega) \geqslant 0$ (for some $\epsilon > 0$, $k(t) > \epsilon$ and $\text{Re } G(j\omega) \geqslant \epsilon$). Moreover, the elements of $\mathcal{G}_T$ which satisfy the above inequality are the only elements of $\mathcal{G}$ which yield (strictly) positive operators KG and GK for all (strictly) positive elements $K \in \mathcal{K}_T$.

Proof: The first part of the theorem follows from Lemma 3.3 if it is proved for KG. But by Lemmas 3.3 and 3.4

$$\langle x, KGx \rangle = \langle x, K^{1/2}GK^{1/2}x \rangle.$$

Since $K^{1/2} \in \mathcal{K}$, it is self-adjoint, and thus

$$\langle x, KGx \rangle = \langle K^{1/2}x, GK^{1/2}x \rangle,$$

which is positive by Theorem 3.1. To prove the strict positivity condition, write KG as $KG = (K - \epsilon I)G + \epsilon G$ and apply the previous part of this theorem and Theorem 3.1. For the converse part of the theorem, assume first that Re $G(j\omega) < 0$ for all ω in a set of positive measure. The result then follows by contradiction if one chooses $K = I$. Assume next that Re $G(j\omega) \geqslant 0$ for almost all ω, but that

$$G(j(\omega + 2\pi T^{-1})) - G(j\omega) \neq 0$$

for all ω in a set of positive measure, say Ω. This part of the theorem is then proved by choosing particular functions for $k(t)$ and $x(t)$ which lead to $\langle x, KGx \rangle < 0$. For simplicity assume that

$$\text{Re } (G(j\omega) - G(j(\omega + 2\pi T^{-1}))) < 0$$

on the set Ω. (A similar argument holds for the other cases). Then there exists an $\epsilon > 0$ such that Re $(G(j\omega) - G(j(\omega + 2\pi T^{-1}))) \leqslant -\epsilon$ for all $\omega \in \Omega'$ with $\Omega' \subset \Omega$ a set of positive measure. Let Ω_n'' be a subset of $[n2\pi T^{-1}, (n + 1)2\pi T^{-1}] \cap \Omega'$ which is of positive measure (such a subset exists for some n, since Ω' is of positive measure). Let $\Omega_n'' + k2\pi T^{-1}$ denote the set of points $x \in R$ such that $x - k2\pi T^{-1} \in \Omega_n''$. Choose $\hat{X}(j\omega) = 1$ for $\omega \in \Omega_n'' + k2\pi T^{-1}$, $k \in I$, and $|k| \leqslant N$, and $\hat{X}(j\omega) = 0$ otherwise, and choose $k(t) = 1 - \cos 2\pi T^{-1}t$. Clearly, $k(t) \geqslant 0$ and the K corresponding to $k(t)$ belongs to K_T. Let $y = KGx$. Then

$$\hat{Y}(j\omega) = G(j\omega)\hat{X}(j\omega) - \tfrac{1}{2}G(j(\omega + 2\pi T^{-1}))\hat{X}(j(\omega + 2\pi T^{-1}))$$

$$- \tfrac{1}{2}G(j(\omega - 2\pi T^{-1}))\hat{X}(j(\omega - 2\pi T^{-1})).$$

A simple calculation shows that the inner product $\langle x, KGx \rangle$ becomes

$$M + (N/\pi) \text{ Re } (G(j\omega) - G(j(\omega + 2\pi T^{-1})))\mu(\Omega_n''),$$

with M a number independent of N, and $\mu(\Omega_n'')$ the Lebesgue measure of Ω_n''. Thus $\langle x, KGx \rangle$ can be made negative by choosing N sufficiently large.[7]

---

[7] Similar positive operators involving a time-varying gain have appeared in the literature. In particular the positive operator obtained by Gruber and Willems (Ref. 3), and in its full generality by Freedman and Zames (Ref. 4), appears to be especially interesting. By restricting the derivative of $k(t)$, they obtain a class of operators which can then be composed with K without destroying positivity.

3.5 Positive Operators with Monotone or Odd-Monotone Nonlinearities

In this section an answer is given to the following question: What is the most general linear operator that when composed with a monotone nondecreasing (or an odd-monotone nondecreasing) nonlinearity yields a positive operator? The answer to this question is the solution to a problem which has been studied by many previous researchers.[8]

The preliminary result obtained in this section constitutes a considerable extension of a classical rearrangement inequality. This inequality then forms the basis from which the positive operators are derived. It is felt that these rearrangement inequalities are of intrinsic importance and that they will be useful in other areas of system theory. For various technical reasons, the discussion is mainly concerned with sequences. With some modifications, similar results can be obtained for the continuous case. Hence, these will merely be stated without proof.

3.5.1 Generalizations of a Classical Rearrangement Inequality

Chapter 10 of Hardy, Littlewood, and Polya's classic book on inequalities (Ref. 15, p. 277) is devoted to questions relating the inner products of real-valued similarly-ordered sequences to the inner products of rearranged sequences. The simplest result given there states that if $x_1 \geqslant x_2 \geqslant \cdots \geqslant x_n$ and $y_1 \geqslant y_2 \geqslant \cdots \geqslant y_n$ and if $y_{\pi(1)}, y_{\pi(2)}, \ldots, y_{\pi(n)}$ is any rearrangement of the y-sequence then

$$\sum_{k=1}^{n} x_k y_k \geqslant \sum_{k=1}^{n} x_k y_{\pi(k)}$$

The informal explanation of this fact given there is that, given a lever arm with hooks at distances $x_1, x_2, \ldots, x_n$ from a pivot and weights $y_1, y_2, \ldots, y_n$ to hang on the hooks, the largest moment is obtained by hanging the largest weight on the farthest hook, the next largest weight on the next most distant hook, etc.

This result has an interpretation in terms of positive operators. Suppose that f is a function from R into itself, and denote by x and Fx

[8] In particular it is the problem studied by Page (Ref. 5), Pantell (Ref. 6), and Black (Ref. 7) in connection with frequency-power formulas and it plays a central role in the determination of stability criteria for feedback systems with a monotone or an odd-monotone nonlinearity in the feedback loop. In the latter context it has been treated by Brockett and Willems (Ref. 8), Narendra and Neumann (Ref. 9), Zames (Ref. 10), O'Shea (Refs. 11, 12), O'Shea and Younis (Ref. 13), Zames and Falb (Ref. 14), and others.

the n-vectors whose components are respectively $x_1, x_2, \ldots, x_n$ and $f(x_1), f(x_2), \ldots, f(x_n)$. Then in the language of positive operators the Hardy, Littlewood, and Polya rearrangement theorem states that the operator N on R^n defined by $Nx \triangleq (I - P)Fx$ is positive if I is the identity matrix, P is any permutation matrix, and f is monotone nondecreasing.

It will be shown that this result together with a result of Birkhoff on the decomposition of doubly stochastic matrices permits the derivation of a number of interesting positivity conditions for a class of operators.[9] The results thus represent a test for checking the positivity of a class of nonquadratic forms parallel to the Sylvester test for checking the positive definiteness of a symmetric matrix.

Definitions: Two sequences of real numbers $\{x_1, x_2, \ldots x_n\}$ and $\{y_1, y_2, \ldots y_n\}$ are said to be *similarly ordered* if the inequality $x_k < x_l$ implies that $y_k \leqslant y_l$. Thus two sequences are similarly ordered if and only if they can be rearranged in such a way that the resulting sequences are both monotone nondecreasing; i.e., there exists a permutation $\pi(k)$ of the first n integers ($\pi(k)$ takes on each of the values $1, 2, \ldots, n$ just once as k varies through the values $1, 2, \ldots, n$) such that *both* the sequences $\{x_{\pi(1)}, x_{\pi(2)}, \ldots, x_{\pi(n)}\}$ and $\{y_{\pi(1)}, y_{\pi(2)}, \ldots, y_{\pi(n)}\}$ are monotone nondecreasing. Two sequences are said to be *unbiased* if $x_k y_k \geqslant 0$ for $1 \leqslant k \leqslant n$. Clearly two sequences are similarly ordered and unbiased if and only if the augmented sequences $\{x_1, x_2, \ldots, x_n, x_{n+1}\}$ and $\{y_1, y_2, \ldots, y_n, y_{n+1}\}$ with $x_{n+1} = y_{n+1} = 0$ are similarly ordered. Two sequences are said to be *similarly ordered and symmetric* if they are unbiased and if the sequences $\{|x_1|, |x_2|, \ldots, |x_n|\}$ and $\{|y_1|, |y_2|, \ldots, |y_n|\}$ are similarly ordered.

As an example, let $f(\sigma)$ be a mapping from the real line into itself, and consider the sequences $\{x_1, x_2, \ldots x_n\}$ and $\{f(x_1), f(x_2), \ldots, f(x_n)\}$. These two sequences will be similarly ordered for all sequences $\{x_1, x_2, \ldots, x_n\}$ if and only if $f(\sigma)$ is a *monotone nondecreasing* function of σ, i.e., if for all σ_1 and σ_2, $(\sigma_1 - \sigma_2)(f(\sigma_1) - f(\sigma_2)) \geqslant 0$. They will be unbiased if and only if $f(\sigma)$ is a *first and third quadrant function;* i.e., if for all σ, $\sigma f(\sigma) \geqslant 0$. They will be similarly ordered and symmetric if and only if $f(\sigma)$ is an *odd monotone nondecreasing function* of σ, i.e., if $f(\sigma)$ is monotone nondecreasing and $f(\sigma) = -f(-\sigma)$ for all σ.

---

[9] The Hardy, Littlewood, and Polya rearrangement inequality leads rather easily to a proof of the fact that the cross correlation of the input and the output to a monotone nondecreasing nonlinearity attains its maximum value at the origin. This approach has in fact been used by Prosser (Ref. 16) and Black (Ref. 7).

Definitions: A real $(n \times n)$ matrix $M = (m_{kl})$ is said to be *doubly hyperdominant*[10] *with zero excess* if $m_{kl} \leqslant 0$ for $k \neq l$, and if $\sum_{k=1}^{n} m_{kl} = \sum_{l=1}^{n} m_{kl} = 0$ for all k, l. It is said to be *doubly hyperdominant* if $m_{kl} \leqslant 0$ for $k \neq l$, and if $\sum_{k=1}^{n} m_{kl} \geqslant 0$ and $\sum_{l=1}^{n} m_{kl} \geqslant 0$ for all k, l. An $(n \times n)$ matrix M is said to be *doubly dominant* if $m_{ll} \geqslant \sum_{k=1, k \neq l}^{n} |m_{kl}|$ and $m_{kk} \geqslant \sum_{l=1, l \neq k}^{n} |m_{kl}|$. It is clear that all of the classes of matrices introduced above are subclasses of the class of all matrices whose symmetric part is nonnegative definite and that every doubly hyperdominant matrix is doubly dominant.

Two other classes of matrices that will be used in the sequel and have received ample attention in the past are now defined.

A $(n \times n)$ matrix M is said to be *doubly stochastic* if it is nonnegative (i.e., $m_{kl} \geqslant 0$ for all k, l) and if each row and column sums to one. A $(n \times n)$ matrix is said to be a *permutation* matrix if every row and column contains $n - 1$ zero elements and an element which equals 1. The relation between the class of doubly stochastic matrices and permutation matrices is given in the following lemma due to Birkhoff.

LEMMA 3.5

The set of all doubly stochastic matrices forms a convex polyhedron with the permutation matrices as vertices; i.e., if M is a doubly stochastic matrix then $M = \sum_{i=1}^{N} \alpha_i P_i$ with $\alpha_i \geqslant 0$, $\sum_{i=1}^{N} \alpha_i = 1$, and P_i a permutation matrix. This decomposition need not be unique.

A proof of Lemma 3.5 can be found in most books on matrix theory (see, e.g., Ref. 20, p. 97).

Theorem 3.7 states the main result of this section. As mentioned before, it constitutes a considerable generalization of a classical rearrangement inequality due to Hardy, Littlewood, and Polya. This inequality is stated in Lemma 3.6.

LEMMA 3.6

Let $\{x_1, x_2, \ldots, x_n\}$ and $\{y_1, y_2, \ldots, y_n\}$ be two similarly ordered sequences, and let $\pi(k)$ be a permutation of the first n integers. Then $\sum_{k=1}^{n} x_k y_k \geqslant \sum_{k=1}^{n} x_k y_{\pi(k)}$.

---

[10] The term "dominant" is standard. "Hyperdominance" is prevalent, at least in the electrical network synthesis literature. The term "doubly" is used by analogy with "doubly stochastic" where a property of a matrix also holds for its transpose. Beyond this the nomenclature originates with the author. Symmetric hyperdominant matrices are sometimes called *Stieltjes Matrices*. For an interesting application of these matrices to the resistive N-part problem, see Reference 17. For some general theorems on matrices of the type under consideration, see References 18, 19.

A simple proof of Lemma 3.6 can be found in Hardy, Littlewood, and Polya's book (Ref. 15, p. 277). A convincing plausibility argument is given in the introduction to this section.

THEOREM 3.7

A necessary and sufficient condition for the bilinear form $\sum_{k,l=1}^{n} m_{kl}x_{k}y_{l}$ to be nonnegative for all similarly ordered sequences $\{x_1, x_2, \ldots x_n\}$ and $\{y_1, y_2, \ldots y_n\}$ is that the matrix $M = (m_{kl})$ be doubly hyperdominant with zero excess.

Proof of sufficiency: Let M be a doubly hyperdominant matrix with zero excess and let r be any positive number such that $r \geqslant m_{kl}$ for all k, l. Clearly $M = r[I - (1/r)(rI - M)]$. Since however $(1/r)(rI - M)$ is a doubly stochastic matrix, it can, by Lemma 3.5, be decomposed as $\sum_{i=1}^{N} \alpha_i P_i$ with $\alpha_i \geqslant 0$, $\sum_{i=0}^{N} \alpha_i = 1$, and the P_i's permutation matrices. Thus M can be written as $M = \sum_{i=1}^{N} \beta_i(I - P_i)$ with $\beta_i \geqslant 0$. This decomposition of doubly hyperdominant matrices with zero excess shows that it is enough to prove the sufficiency part of Theorem 3.7 for the matrices $I - P_i$. This, however, is precisely what is stated in Lemma 3.6.

Proof of necessity: The matrix M may fail to be doubly hyperdominant with zero excess because $m_{kl} > 0$ for some $k \neq l$ in which case the sequences with $n - 1$ zero elements except $+1$ and -1 in respectively the kth and lth places lead to $\sum_{k,l=1}^{n} m_{kl}x_{k}y_{l} = -m_{kl} < 0$. Assume next that the matrix M fails to be doubly hyperdominant with zero excess because $\sum_{k=1}^{n} m_{kl} \neq 0$ for some l (a similar argument holds if $\sum_{l=1}^{n} m_{kl} \neq 0$ for some k), and consider the similarly ordered sequences $\{1, \ldots, 1, 1 + \epsilon, 1, \ldots, 1\}$ and $\{0, \ldots, 0, \epsilon^{-1}, 0, \ldots, 0\}$ with $\epsilon \neq 0$, and the elements $1 + \epsilon$ and ϵ^{-1} in the lth place. This leads to $\sum_{k,l=1} m_{kl}x_{k}y_{l} = \epsilon^{-1}\sum_{k=1}^{n} m_{kl} + m_{ll}$. By taking ϵ sufficiently small and of an appropriate sign, $\sum_{k,l=1}^{n} m_{kl}x_{k}y_{l}$ can then indeed be made negative.

The following two theorems are generalizations of Theorem 3.7 to similarly ordered unbiased and to similarly ordered symmetric sequences.

THEOREM 3.8

A necessary and sufficient condition for the bilinear form $\sum_{k,l=1}^{n} m_{kl}x_{k}y_{l}$ to be nonnegative for all similarly ordered unbiased sequences $\{x_1, x_2, \ldots, x_n\}$ and $\{y_1, y_2, \ldots, y_n\}$ is that the matrix $M = (m_{kl})$ be doubly hyperdominant.

Proof of sufficiency: Let M be a doubly hyperdominant matrix and define $m_{k,n+1} = -\sum_{l=1}^{n} m_{kl}$, $m_{n+1,l} = -\sum_{k=1}^{n} m_{kl}$ for k, $l \leqslant n$, and $m_{n+1,n+1} = \sum_{k,l=1}^{n} m_{kl}$. Then taking $x_{n+1} = y_{n+1} = 0$ it follows from Theorem 3.7 that $\sum_{k,l=1}^{\tilde{n}} m_{kl} x_k y_l = \sum_{k,l=1}^{\tilde{n}} m_{kl} x_k y_l \geqslant 0$ since the augmented $(n+1 \times n+1)$ matrix $M_* = (m_{kl})$, $k, l = 1, 2, \ldots, n+1$ is doubly hyperdominant with zero excess and since the sequences $\{x_1, x_2, \ldots, x_n, x_{n+1}\}$ and $\{y_1, y_2, \ldots, y_n, y_{n+1}\}$ with $x_{n+1} = y_{n+1} = 0$ are similarly ordered.

Proof of necessity: The same sequences as in Theorem 3.7 can be used if the matrix M fails to be doubly hyperdominant because $m_{kl} > 0$ for some $k \neq l$. Assume next that the matrix M fails to be doubly hyperdominant because $\sum_{k=1}^{n} m_{kl} < 0$ for some l (a similar argument holds if $\sum_{l=1}^{n} m_{kl} < 0$ for some k), and consider the sequences used in Theorem 3.7 with the additional restriction that $\epsilon > 0$. Notice that these sequences are then similarly ordered and unbiased. It follows that by taking ϵ sufficiently small $\sum_{k,l=1}^{n} m_{kl} x_k y_l = \epsilon^{-1} \sum_{k=1}^{n} m_{kl} + m_{ll}$ can thus be made negative.

THEOREM 3.9

A necessary and sufficient condition for the bilinear form $\sum_{k,l=1}^{n} m_{kl} x_k y_l$ to be nonnegative for all similarly ordered symmetric sequences $\{x_1, x_2, \ldots, x_n\}$ and $\{y_1, y_2, \ldots, y_n\}$ is that the matrix $M = (m_{kl})$ be doubly dominant.

Proof of sufficiency: Let M be a doubly dominant matrix. Clearly,

$$\sum_{\substack{k,l=1 \\ k=l}}^{n} m_{kl} x_k y_l \geqslant \sum_{\substack{k,l=1 \\ k=l}}^{n} m_{kl} |x_k| |y_l| - \sum_{\substack{k,l=1 \\ k \neq l}}^{n} |m_{kl}| |x_k| |y_l|.$$

The right-hand side of the this inequality is nonnegative by Theorem 3.8, since the matrix $M^* = (m_{kl}^*)$ with $m_{kl}^* = m_{kl}$ when $k = l$ and $m_{kl}^* = -|m_{kl}|$ when $k \neq l$ is doubly hyperdominant and since the sequences $\{|x_1|, |x_2|, \ldots, |x_n|\}$ and $\{|y_1|, |y_2|, \ldots, |y_n|\}$ are similarly ordered and unbiased. This implies that $\sum_{k,l=1}^{n} m_{kl} x_k y_l \geqslant 0$.

Proof of necessity: Assume that the matrix M fails to be doubly dominant because $m_{ll} - \sum_{k=1,k \neq l}^{n} |m_{kl}| < 0$ for some l (an analogous argument holds if $m_{kk} - \sum_{l=1,l \neq k}^{n} |m_{kl}| < 0$ for some k), and consider the sequences $\{-\text{sgn } m_{1l}, \ldots, -\text{sgn } m_{l-1,l}, 1 + \epsilon, -\text{sgn } m_{l+1,l}, \ldots, -\text{sgn } m_{nl}\}$ and $\{0, \ldots, 0, \epsilon^{-1}, 0, \ldots, 0\}$ with $\text{sgn } \alpha = \alpha/|\alpha|$ if $\alpha \neq 0$, $\text{sgn } 0 = 0$, $\epsilon > 0$, and the elements $1 + \epsilon$ and ϵ^{-1} in the lth places. These sequences are similarly ordered and symmetric and lead to

$\sum_{k,l=1}^{n} m_{kl}x_{k}y_{l} = \epsilon^{-1}(m_{ll} - \sum_{k=1,k\neq l}^{n} |m_{kl}|) + m_{ll}$, which, by taking ϵ sufficiently small, yields $\sum_{k,l=1}^{n} m_{kl}x_{k}y_{l} < 0$.

Let f be a mapping from R into R and denote by F the mapping from R_n into itself which takes the element col $(x_1, x_2, \ldots, x_n)$ into col $(f(x_1), f(x_2), \ldots, f(x_n))$. Then in terms of positive operators, Theorems 3.7–3.9 become

THEOREM 3.10

Let M be an $(n \times n)$ matrix and let f be one of the following:

1. A monotone nondecreasing function.
2. A monotone nondecreasing first and third quadrant function.
3. An odd-monotone nondecreasing function.

Then MF is a positive operator on R^n for all mappings f satisfying one of the conditions 1–3 if and only if M is, respectively,

1. A doubly hyperdominant matrix with zero excess.
2. A doubly hyperdominant matrix.
3. A doubly dominant matrix.

3.5.2 Extension to l_2-summable Sequences

In this section l_p is taken over the field of real numbers unless otherwise mentioned.

Definitions: Let $\mathcal{L}(l_2,l_2)$ denote all bounded linear transformations from l_2 into itself. Let $R \in \mathcal{L}(l_2,l_2)$. Then R determines an array of real numbers $\{r_{kl}\}$, $k, l \in I$, such that $y = Rx$ is defined by $y_k = \sum_{k=-\infty}^{+\infty} r_{kl}x_l$ for $x = \{x_k\}$ and $y = \{y_k\}$, $k \in I$ (Reference 21, p. 50). This infinite sum exists for all $x \in l_2$ and the resulting sequence belongs to l_2. A standard result in the theory of bounded linear operators in Hilbert space (Ref. 21, p. 52) states that the array $\{r_{kl}^*\}$, $k, l \in I$ corresponding to the adjoint R^*, is given by $r_{kl}^* = r_{lk}$ for all $l, k \in I$. It is *not* known what arrays determine elements of $\mathcal{L}(l_2,l_2)$. The following lemma, however, covers a wide class.

LEMMA 3.7

Let the array $\{r_{kl}\}$, $k, l \in I$, be such that the sequences $\{r_{kl}\}$ belong to l_1 for fixed k and l, uniformly in k and l; i.e., there exists an $M < \infty$ such that $\sum_{l=-\infty}^{+\infty} |r_{kl}| \leqslant M$ and $\sum_{k=-\infty}^{+\infty} |r_{kl}| \leqslant M$. Then $\{r_{kl}\}$ determines an element R of $\mathcal{L}(l_2,l_2)$ and $\|R\| \leqslant M$.

Proof: The Schwartz inequality and Fubini's Theorem for sequences (Ref. 27, p. 245) yield the following inequalities:

$$\left(\sum_{k=-\infty}^{+\infty}\left|\sum_{l=-\infty}^{+\infty}r_{kl}x_l\right|^2\right)^{1/2} \leqslant \left[\sum_{k=-\infty}^{+\infty}\left(\sum_{l=-\infty}^{+\infty}|r_{kl}||x_l|\right)^2\right]^{1/2}$$

$$\leqslant \left[\sum_{k=-\infty}^{+\infty}\left(\sum_{l=-\infty}^{+\infty}|r_{kl}|\right)\left(\sum_{l=-\infty}^{+\infty}|r_{kl}||x_l|^2\right)\right]^{1/2}$$

$$\leqslant M\left(\sum_{l=-\infty}^{+\infty}|x_l|^2\right)^{1/2}$$

which proves the lemma.

In what follows an important role will be played by some particular elements of $\mathscr{L}(l_2,l_2)$ and some particular sequences which will now be introduced.

Definitions: The definitions of *similarly ordered, similarly ordered unbiased,* and *similarly ordered symmetric* infinite sequences are completely analogous to those for finite sequences and will not be repeated here. It is easy to show that two sequences in l_2 are similarly ordered if and only if they are similarly ordered *and* unbiased. Let M be an element of $\mathscr{L}(l_2,l_2)$, and let $\{m_{kl}\}$, $k, l \in I$ be the associated array. M is said to be *doubly hyperdominant* if $m_{kl} \leqslant 0$ for $k \neq l$ and if $\sum_{k=-\infty}^{+\infty} m_{kl}$ and $\sum_{l=-\infty}^{+\infty} m_{kl}$ exist and are nonnegative for all l and k. M is said to be *doubly dominant* if

$$m_{ll} \geqslant \sum_{\substack{k=-\infty \\ k \neq l}}^{+\infty} |m_{kl}| \quad \text{and} \quad m_{kk} \geqslant \sum_{\substack{l=-\infty \\ l \neq k}}^{+\infty} |m_{kl}|.$$

It is clear from Lemma 3.7 that if an array or real numbers $\{m_{kl}\}$, $k, l \in I$, satisfies the doubly dominance condition and if the sequence $\{m_{kk}\} \in l_\infty$, then $\{m_{kl}\}$ determines an element M of $\mathscr{L}(l_2,l_2)$ with $\|M\| \leqslant 2 \sup_{k \in I} m_{kk}$. Thus it is a relatively simple matter to check whether or not an element of $\mathscr{L}(l_2,l_2)$ is doubly hyperdominant or doubly dominant.

The following extensions of Theorems 3.8 and 3.9 hold:

THEOREM 3.11

Let M be an element of $\mathscr{L}(l_2,l_2)$. Then a necessary and sufficient condition for the inner product $\langle x, My \rangle_{l_2}$ to be nonnegative for all similarly ordered unbiased l_2-sequences x and y (similarly ordered symmetric l_2-sequences x and y) is that M be doubly hyperdominant (doubly dominant).

Proof: It is clear that all finite subsequences of x and y are similarly ordered and unbiased or similarly ordered and symmetric. Hence, by Theorems 3.8 and 3.9 all finite truncations of the infinite sum in the inner product $\langle x, My \rangle$ yield a nonnegative number. Thus the limit, since it exists, is also nonnegative.

Of particular interest are the arrays $\{r_{kl}\}$, k, $l \in I$, for which the entries depend on the difference of the indices k and l only. These arrays are said to be of the *Toeplitz type* and have been intensively studied in classical analysis (Ref. 23). It follows from Lemma 3.7 that if the array $\{r_{kl} = r_{k-l}\}$, k, $l \in I$, is of the Toeplitz type then it determines an element of $\mathscr{L}(l_2, l_2)$ if $\{r_k\}$, $k \in I$, belongs to l_1. (In fact, the elements of $\mathscr{L}(l_2, l_2)$ for which the associated array is of the Toeplitz type stand in one-to-one correspondence to all l_2-summable sequences whose limit-in-the-mean z-transform belongs to L_∞ for $|z| = 1$.) An element of $\mathscr{L}(l_2, l_2)$ is said to be of the *Toeplitz type* if the associated array is of the Toeplitz type. An element R of $\mathscr{L}(l_2, l_2)$ which is of the Toeplitz type thus determines a sequence $\{r_k\}$, $k \in I$, with $\{r_k\} \in l_2$ and whose limit-in-the-mean z-transform belongs to L_∞ for $|z| = 1$. The importance of these linear transformations stems from the fact that they define convolution operators with a time-invariant kernel and are therefore associated with time-invariant systems.[11]

Definitions: A sequence of real numbers $\{a_k\}$, $k \in I$, is said to be *hyperdominant* if $\{a_k\} \in l_1$, if $a_k \leqslant 0$ for all $k \neq 0$, and if $\sum_{k=-\infty}^{+\infty} a_k \geqslant 0$. It is said to be *dominant* if $\{a_k\} \in l_1$, and if $2a_0 \geqslant \sum_{k=-\infty}^{+\infty} |a_k|$.

THEOREM 3.12

Let M be an element of $\mathscr{L}(l_2, l_2)$ which is of the Toeplitz type. Then a necessary and sufficient condition for the inner product $\langle x, My \rangle$ to be nonnegative for all similarly ordered unbiased l_2-sequences x and y (similarly ordered symmetric l_2-sequences x and y) is that the sequence $\{m_k\}$, $k \in I$, which is determined by M be hyperdominant (dominant).

Theorem 3.12 is a special case of Theorem 3.11. Theorems 3.11 and 3.12 have an obvious interpretation in terms of positive operators. Moreover, Theorem 3.12 yields some simple properties of the input

---

[11] A definition of time invariance for operators defined on the continuous real line has been given in Chapter 2. The extension of the definition to operators defined with the integers as the time-interval of definition is obvious. It is then a simple matter to verify that an element $R \in \mathscr{L}(l_2, l_2)$ is of the Toeplitz type if and only if R is time invariant.

and the output spectra to (odd) monotone nondecreasing nonlinearities. This is stated explicitly in Theorem 3.13.

Definitions: Let $\mathscr{A}$ denote the class of operators from l_2 into itself, each element of which has associated with it a function $A(z)$, with $A(z) \in L_\infty$ for $|z| = 1$, $A(\bar{z}) = \bar{A}(z)$, and which maps the element x of l_2 as follows: let $\hat{X}$ denote the limit-in-the-mean z-transform of x. The image of x is the sequence y whose limit-in-the-mean z-transform equals $A(z)\hat{X}(z)$.

Let $\mathscr{F}$ denote the class of operators from l_2 into itself, each element of which has associated with it a function $f(\sigma)$ from R into itself, which satisfies the inequality $|f(\sigma)| \leqslant M |\sigma|$ for some $M < \infty$ and all $\sigma \in R$ and which maps the sequence $x = \{x_k\}$, $k \in I$, of l_2 into the sequence $y = \{y_k\}$ with $y_k \triangleq f(x_k)$.

It is a simple matter to verify that these operators are indeed well defined, i.e., that they map l_2 into itself. The class $\mathscr{A}$ stands in one-to-one correspondence with all l_2-sequences whose limit-in-the-mean z-transform belongs to L_∞ for $|z| = 1$. Moreover if $\{a_k\} \in l_2$ and $A(z) \in L_\infty$ for $|z| = 1$ are such a sequence and its limit-in-the-mean z-transform, then the element of $\mathscr{A}$ which has the function $A(z)$ associated with it maps l_2 into itself according to the convolution $y_k = \sum_{l=-\infty}^{+\infty} a_{k-l} x_l$.

THEOREM 3.13

Let $A \in \mathscr{A}$ and $F \in \mathscr{F}$. Then AF is a positive operator on l_2 if the following statements are true:

1. The f corresponding to F is a (odd) monotone nondecreasing first and third quadrant function.
2. The inverse z-transform of $A(z)$ is (dominant) hyperdominant.

Moreover, the elements of $\mathscr{A}$ satisfying statement 2 are the most general elements of $\mathscr{A}$ which yield a nonnegative operator AF on l_2, for any $F \in \mathscr{F}$ satisfying statement 1. Finally AF is a strictly positive operator on l_2 if $A - \epsilon I$ and $F - \epsilon I$ satisfy statements 1 and 2 for some $\epsilon > 0$.

The proof of Theorem 3.13 follows from Theorem 3.12. Theorem 3.13 states that if $\hat{X}(z)$ and $\hat{Y}(z)$ are the limit-in-the-mean z-transforms of the input and the output of a (odd) monotone nondecreasing nonlinearity then

$$\oint_{|z|=1} A(z)\,\hat{X}(z)\,\hat{Y}(\bar{z})\,dz \geqslant 0$$

for any $A(z)$ which is the z-transform of a (dominant) hyperdominant sequence.

3.6 Frequency-Power Relations for Nonlinear Resistors

In this section a class of positive operators formed by the composition of a linear time-invariant convolution operator and a (odd) monotone nondecreasing nonlinearity will be derived. The analysis is done for operators on L_2, but the results are also stated for almost periodic functions — thus placing the positive operators obtained in this section in the context of the classical frequency-power relations for nonlinear resistors.

Definitions: Let $\mathcal{M}$ denote the class of operators from L_2 into itself each element of which belongs to the class $\widetilde{\mathcal{F}}$ and for which the associated function f is monotone nondecreasing; i.e., $(\sigma_1 - \sigma_2)(f(\sigma_1) - f(\sigma_2)) \geqslant 0$ for all $\sigma_1, \sigma_2 \in R$. Let $\mathcal{S}$ denote the class of operators from L_2 into itself each element of which belongs to $\mathcal{M}$ and for which the associated function f is in addition an *odd* function; i.e., $f(\sigma) = -f(-\sigma)$ for all $\sigma \in R$.

Let $x_1, x_2 \in L_2$. Then $x_2(t + \tau) \in L_2$ for all $\tau \in R$, and $\|x_2(t + \tau)\| = \|x_2(t)\|$. The *crosscorrelation* function of x_1 and x_2 is defined as the function $R_{x_1 x_2}(\tau) = \langle x_1(t), x_2(t + \tau) \rangle$. Note that the Schwartz inequality yields $|R_{x_1 x_2}(\tau)| \leqslant \|x_1\| \|x_2\|$. Moreover, since the limit-in-the-mean transforms of $x(t)$ and $x(t + \tau)$ are given by $\hat{X}(j\omega)$ and $\hat{X}(j\omega)e^{j\omega\tau}$ respectively, it follows from Parseval's relation that

$$R_{x_1 x_2}(\tau) = \frac{1}{2\pi} \int_{-\infty}^{+\infty} \hat{X}_1(-j\omega)\hat{X}_2(j\omega)e^{j\omega\tau}\,d\omega.$$

The above crosscorrelation-function inequality states in particular that

$$|R_{x_1 x_1}(\tau)| \leqslant R_{x_1 x_1}(0) = \|x_1\|^2.$$

The theorem which follows is a generalization of this well-known property of autocorrelation functions. It states that the crosscorrelation function of x and y attains its maximum at the origin provided x and y are related through a monotone nondecreasing nonlinearity.[12]

---

[12] This property follows rather easily from the Hardy, Littlewood, and Polya rearrangement inequality. This was pointed out by Prosser (Ref. 16) and Black (Ref. 7). The inequality $R_{xy}(0) \geqslant \frac{1}{2}(R_{xy}(t) + R_{xy}(-t))$ is easy to obtain and holds for any x and y which are related through an incrementally positive time-invariant operator. This inequality in turn implies incremental positivity and time invariance. It is not clear, except in the scalar case, precisely what conditions on F are necessary and sufficient to insure that $R_{xy}(0) \geqslant R_{xy}(t)(R_{xy}(0) \geqslant |R_{xy}(t)|)$ for any x and $y = Fx$.

THEOREM 3.14

Let $F \in \mathcal{M}$, $x \in L_2$ and let $y = Fx$. Then $R_{xy}(0) \geqslant R_{xy}(t)$ for all $t \in R$. If F belongs to $\mathcal{S}$, then $R_{xy}(0) \geqslant |R_{xy}(t)|$ for all $t \in R$.

Proof: Let $F(\sigma) = \int_0^\sigma f(x)\,dx$. Then $F(\sigma)$ is a convex function of σ (since the derivative of $F(\sigma)$ exists and is monotone nondecreasing). The convex function inequality (Ref. 24) yields that $(\sigma_1 - \sigma_2)f(\sigma_1) \geqslant F(\sigma_1) - F(\sigma_2)$ for all $\sigma_1, \sigma_2 \in R$. (This inequality can simply be obtained by integrating $f(\sigma) - f(\sigma_1)$ from σ_1 to σ_2.) Taking $\sigma_1 = x(t + \tau)$ and $\sigma_2 = x(t)$ it follows that

$$(x(t) - x(t + \tau))y(t) \geqslant F(x(t)) - F(x(t + \tau)),$$

which after integration yields

$$R_{xy}(0) - R_{xy}(\tau) \geqslant \int_{-\infty}^{+\infty} F(x(t))\,dt - \int_{-\infty}^{+\infty} F(x(t + \tau))\,dt = 0.$$

The integrals on the right-hand side exist since by assumption $F \in \mathcal{M}$ and thus $|f(\sigma)| \leqslant K|\sigma|$ for some K and all $\sigma \in R$, which implies that $|F(\sigma)| \leqslant \frac{1}{2}K|\sigma|^2$ for all $\sigma \in R$. Hence $R_{xy}(0) \geqslant R_{xy}(t)$ for all $F \in \mathcal{M}$ and $t \in R$. If f is in addition odd, then the convex function inequality can be rewritten as $[\sigma_1 - (-\sigma_2)]f(\sigma_1) \geqslant F(\sigma_1) - F(-\sigma_2)$, which, since f is odd, yields that $(\sigma_1 + \sigma_2)f(\sigma_1) \geqslant F(\sigma_1) - F(\sigma_2)$. Exactly the same argument as above then leads to $R_{xy}(0) + R_{xy}(t) \geqslant 0$ for all $t \in R$. Thus, $R_{xy}(0) \geqslant |R_{xy}(t)|$ for all $F \in \mathcal{S}$ and $t \in R$.

Remark: It can be shown (Ref. 25) that Theorem 3.14 is also sufficient in the sense that if $y = Fx$ for some $F \in \tilde{\mathcal{F}}$ and if $R_{xy}(0) \geqslant R_{xy}(t)$ $(R_{xy}(0) \geqslant |R_{xy}(t)|)$, for all $x \in L_2$ and $t \in R$, then $F \in \mathcal{M}(\mathcal{S})$.

THEOREM 3.15

Let $F \in \mathcal{M}(\mathcal{S})$ and let $G \in \mathcal{G}$ be determined by the function $G(j\omega)$ given by the Fourier-Stieltjes integral

$$G(j\omega) = 1 - \int_{-\infty}^{+\infty} e^{-j\omega\tau}\,dV(\tau)$$

where $V(\tau)$ is any monotone nondecreasing function (any function of bounded variation) of total variation less than or equal to unity. Then GF is a nonnegative operator on L_2.

Proof: Assume first that $F \in \mathcal{M}$. The theorem follows from the previous theorem if it is noted that $R_{xy}(0) \geqslant 0$ and that the operator G is defined by the convolution

$$y(t) = (Gx)(t) = x(t) - \int_{-\infty}^{+\infty} x(t - \tau) \, dV(\tau).$$

Indeed, let $y = Fx$. Thus

$$\langle x, Gy \rangle = c^2 R_{xy}(0) + \int_{-\infty}^{+\infty} [R_{xy}(0) - R_{xy}(\tau)] \, dV(\tau),$$

where $c^2 = 1 -$ the total variation of V. Note that the above integrals exist, since R_{xy} is bounded and since V is of bounded total variation. Thus $\langle x, Gy \rangle = \langle x, GFx \rangle \geqslant 0$ by Theorem 3.14. The odd-monotone case is proved in a similar way.

Remark: GF will be a positive operator on L_2 if $F - \epsilon I \in \mathcal{M}(\mathcal{S})$ for some $\epsilon > 0$ and if the total variation of V is strictly less than unity.

THEOREM 3.16

Let F and G satisfy the conditions of Theorem 3.15 and assume that the function f which determines F satisfies a Lipschitz condition on R. Then $(G + a \, d/dt)F$ is a nonnegative operator on S_2^1 for all $\alpha \in R$.

The proof of Theorem 3.16 follows from Theorem 3.4 and 3.15. Theorem 3.16 states that if $\hat{X}$ and $\hat{Y}$ are the limit-in-the-mean transforms of x and $y = Fx$ with x and F as in Theorem 3.16, then

$$\int_{-\infty}^{+\infty} M_1(j\omega) \hat{X}(-j\omega) \hat{Y}(j\omega) \, d\omega \geqslant 0$$

for all functions $M_1(j\omega)$ given by the Fourier-Stieltjes integral

$$M_1(j\omega) = 1 + \alpha j\omega - \int_{-\infty}^{+\infty} e^{-j\omega\tau} \, dV_1(\tau),$$

where $\alpha \in R$ and $V_1(\tau)$ satisfies the conditions of Theorem 3.15.

There is one possible refinement of this result that has, however, no immediate interpretation in terms of positive operators unless additional smoothness assumptions are made on x. Consider the functions of the form[13]

$$M_2(j\omega) = \int_{-\infty}^{+\infty} \frac{1 - e^{j\omega\tau} - j\omega\tau g(\tau)}{\tau^2} \, dV(\tau),$$

[13] Functions of this type have been studied in probability analysis in connection with characteristic functions of spatially homogeneous diffusion processes and infinitely divisible distributions (see e.g. Ref. 36, p. 654 and Ref. 26, p. 541). The reason these functions also appear in this context is roughly the following: the doubly hyperdominant matrices with zero excess are the logarithms of doubly stochastic matrices. Thus by considering Markov processes defined on $(-\infty, +\infty)$ rather than finite Markov chains one naturally obtains the continuous versions of the doubly hyperdominant matrices. The transfer functions of the convolution operators obtained here are thus the characteristic functions of the probability density functions of these diffusion processes.

with $V_2(\tau)$ a monotone nondecreasing function of τ (any function of τ which is a bounded variation over compact sets) such that

$$\int_x^\infty \frac{dV_2(\tau)}{\tau^2} \quad \text{and} \quad \int_{-\infty}^{-x} \frac{dV_2(\tau)}{\tau^2}$$

exist for $x > 0$, and $g(\tau)$ is any bounded real-valued function of τ which is continuous at the origin and with $g(0) = 1$ (it can be shown that under these conditions $M_2(j\omega)$ is well defined). It is then possible to show — using an argument which is completely analogous to the one used above — that the integral

$$\int_{-\infty}^{+\infty} M(j\omega)\, \hat{X}(-j\omega)\, \hat{Y}(j\omega)\, d\omega$$

exists and is nonnegative for any $M(j\omega) = M_1(j\omega) + M_2(j\omega)$ with $M_1(j\omega)$ and $M_2(j\omega)$ of the form given above.

The following simple functions of ω belong to this class (for the monotone case) and are of particular interest (Ref. 26, p. 541):

1. The function

$$M(j\omega) = 1 - \gamma \exp\left(-|\omega|^\tau\right),$$

where γ and τ are real numbers satisfying $0 \leqslant \gamma \leqslant 1$ and $0 \leqslant \tau \leqslant 2$.

2. The function

$$M(j\omega) = 1 - e(\omega),$$

where $e(\omega)$ is any real-valued, nonnegative even function of ω which is convex for $\omega \geqslant 0$ and with $e(0) \leqslant 1$.

3. The functions

$$M(j\omega) = |\omega|^\tau\left[1 + j\delta \tan \frac{\pi\tau}{2}\right] \quad \text{for} \quad \omega \geqslant 0,$$

$$M(-j\omega) = \bar{M}(j\omega) \quad \text{for} \quad \omega \leqslant 0.$$

4. The function

$$M(j\omega) = |\omega|\left[1 + j\delta \ln \frac{|\omega|}{\omega_0}\right] \quad \text{for} \quad \omega \geqslant 0,$$

$$M(-j\omega) = \bar{M}(j\omega) \quad \text{for} \quad \omega \leqslant 0.$$

where τ, δ, and ω_0 are real numbers satisfying $0 \leqslant \tau \leqslant 2$, $\tau \leqslant 1$, $|\delta| \leqslant 1$, and $\omega_0 > 0$.

In the remainder of this section these results are tied in with the frequency-power formulas. Frequency-power formulas are relations between the power inputs and the power outputs of a nonlinear device at various frequencies of the almost periodic input. The discussion will be mainly concerned with nonlinear resistors. Frequency-power relations of the type given here have found application in the design of frequency converters and express fundamental limitations of such devices.[14] The proofs of the relations which follow will not be given since they are completely analogous to the parallel relations obtained above for L_2-functions.

Let x be an almost periodic function of t, and let f be a continuous map from R into itself. It follows then from the smoothness conditions on f that $y(t) = f(x(t))$ is also almost periodic.

Definitions: Let ω_k be a basic frequency common to both $x(t)$ and $y(t)$ and let x_k and y_k be the corresponding Fourier coefficients. Let $\omega_k \geqslant 0$. Then the *complex power* R_k, the *active power* P_k, and the *reactive power* Q_k absorbed by the nonlinearity f at frequency ω_k are defined as

$$R_k \triangleq \tfrac{1}{2}\bar{x}_k y_k \qquad P_k \triangleq \operatorname{Re} R_k \qquad Q_k \triangleq \operatorname{Im} R_k.$$

Frequency-power formulas are relations between the active and reactive powers absorbed by the nonlinear resistor at the different frequencies.

The first relation which can then be obtained is the analogue of Theorem 3.4 and states that if x and $\dot{x}$ are almost periodic functions of t, then

$$\sum_{\omega_k \geqslant 0} \omega_k Q_k = 0,$$

i.e., the weighted sum of the reactive power absorbed at various frequencies is zero. This formula is known as the Manley-Rowe frequency-power formula.

If f is a first and third quadrant function then, in analogy with Theorem 3.3,

$$\sum_{\omega_k \geqslant 0} P_k \geqslant 0.$$

If f is in addition monotone nondecreasing (f is then usually referred to as a *nonlinear resistor*), then the following general frequency-power

---

[14] For a complete treatment of frequency-power formulas see the book by Penfield (Ref. 27). The study of these formulas was initiated by Manley and Rowe (Ref. 28), and the work of Carroll (Ref. 29) and Black (Ref. 7) is particularly appropriate in relation to the results presented here.

relation can be obtained in a straightforward fashion:

$$\text{Re} \sum_{\omega_k \geqslant 0} R_k M(j\omega_k) \geqslant 0.$$

The particular choices of M given in the preceding list lead to the following simple frequency-power formulas:

1. The formula

$$\sum_{\omega_k \geqslant 0} [1 - \gamma \exp(-|\omega_k|^r)]P_k \geqslant 0,$$

where γ and τ are real numbers satisfying $0 \leqslant \gamma \leqslant 1$ and $0 \leqslant \tau \leqslant 2$.

2. The formula

$$\sum_{\omega_k \geqslant 0} (1 - e(\omega_k))P_k \geqslant 0,$$

where $e(\omega)$ is any real-valued, nonnegative, even function of ω which is convex for $\omega \geqslant 0$ and with $e(0) \leqslant 1$.

3. The formulas

$$\sum_{\omega_k \geqslant 0} |\omega_k|^r \left(P_k + Q_k \, \delta \tan \frac{\pi\tau}{2}\right) \geqslant 0$$

and

$$\sum_{\omega_k \geqslant 0} \omega_k \left(P_k + Q_k \, \delta \ln \frac{\omega_k}{\omega_0}\right) \geqslant 0,$$

where τ, δ and ω_0 are real numbers satisfying $0 \leqslant \tau \leqslant 2$, $\tau \neq 1$, $|\delta| \leqslant 1$, and $\omega_0 > 0$.

Remark 1: For nonlinear capacitors with voltage versus charge characteristic $v = f(q)$ where f satisfies the same assumptions as above, analogous frequency-power formulas can be obtained with R_k replaced by jR_k/ω_k. The same is true for nonlinear inductors with current versus flux characteristic $i = f(\Phi)$ with R_k replaced by $R_k/j\omega_k$.

Remark 2: An important refinement of the frequency-power relations can be obtained if the input $x(t)$ to the nonlinearity f is assumed to be quasi-periodic, i.e., if it is assumed that the basic frequencies ω_k consist of linear combinations of N fundamental frequencies. Thus, let $\omega_1, \omega_2, \ldots, \omega_N$ be N given real nonnegative numbers, let $n_1, n_2, \ldots, n_N$ be integers, and assume that the basic frequencies in x are of the form $\omega_{n_1 n_2 \ldots n_N} = \sum_{k=1}^{N} n_i \omega_i$. It can then be shown that the basic frequencies of y are of the same type *and* that the Fourier coefficient in the output y at frequency $\omega_{n_1 n_2 \ldots n_N}$ depends on the nonlinearity f, the set of integers $\{n_i\}$ for $i = 1, 2, \ldots, N$, and the set of Fourier coefficients in x, $\{x_{n_1 n_2 \ldots n_N}\}$, $n_1, n_2, \ldots, n_N \in I$, but is *independent* of $\omega_1, \omega_2, \ldots, \omega_N$.[15] Moreover, since

$$\sum_{n_i \geqslant 0} \omega_{n_1 n_2 \ldots n_N} Q_{n_1 n_2 \ldots n_N} = 0,$$

it immediately follows that *each factor* of the different ω_i's in the above equality must be zero. Hence,

$$\sum_{n_i \geqslant 0} n_1 Q_{n_1 n_2 \ldots n_N} = 0$$

$$\cdot$$
$$\cdot$$
$$\cdot$$

$$\sum_{n_i \geqslant 0} n_N Q_{n_1 n_2 \ldots n_N} = 0.$$

In a similar manner one can then argue that with M as before,

$$\text{Re} \sum_{n_i \geqslant 0} R_{n_1 n_2 \ldots n_N} M(j(n_1 \xi_1 + n_2 \xi_2 + \ldots + n_N \xi_N)) \geqslant 0$$

for any real numbers $\xi_1, \xi_2, \ldots, \xi_N$.

3.7 Factorization of Operators

In many problems in system theory, e.g., in stability theory, in optimal control theory, and in prediction theory, there is particular interest in causal operators. For instance, in network synthesis it is clear that a synthesis procedure for passive nonlinear networks will require the operator defining the input-output relation to be both positive and causal. The importance to stability theory of generating positive operators that are also causal will become more apparent later. In this section some techniques for generating a causal positive operator from an arbitrary positive operator are developed. The basic idea is simple and is expressed in the next theorem. The definition of a causal operator has been given in Section 2.4.

THEOREM 3.17

Let F be an (incrementally) positive operator on W with W a Hilbert space of functions defined on the time-interval of definition S and satisfying the usual properties with respect to truncations. Assume that F

---

[15] These facts are obvious when the function f is a power law or a linear combination of power laws. The extension to arbitrary functions f then follows readily by approximation.

admits a factorization $F = F_1 F_2$ on W with F_2 causal on W and F_1 an invertible bounded linear operator on W with $(F_1^*)^{-1}$ causal on W. Then $F_2(F_1^*)^{-1}$ is (incrementally) positive and causal on W.

Proof: Let $x \in W$. Then

$$\langle x, F_2(F_1^*)^{-1}x \rangle = \langle F_1^*(F_1^*)^{-1}x, F_2(F_1^*)^{-1}x \rangle$$
$$= \langle (F_1^*)^{-1}x, F_1 F_2(F_1^*)^{-1}x \rangle \geqslant 0.$$

Hence $F_2(F_1^*)^{-1}$ is positive on W. Incremental positivity is proved in a similar way, and causality of $F_2(F_1^*)^{-1}$ follows since F_2 and $(F_1^*)^{-1}$ are causal by assumption.

This theorem and the resulting possibility of generating a causal positive operator from a noncausal positive operator show the importance of obtaining sufficient conditions for a factorization as required in the theorem to be possible. Similar problems have received a great deal of attention in classical prediction theory, in the theory of linear integral equations, and in probability theory.[16] The existing results deal almost exclusively with linear time-invariant convolution-type operators in Hilbert spaces, and most of the analysis uses the fact that these operators are commutative in an essential way. The operators considered here, however, need not have this property. The factorization theorem obtained in this section is felt to be of great interest in its own right. It applies in particular to linear operators whose kernel might be time-varying and which need therefore not be commutative.

All the nonlinear positive operators obtained in this chapter are compositions of a linear operator and a memoryless (nonlinear) operator. The problem of generating a causal positive operator is thus by virtue of Theorem 3.17 reduced to the factorization of a linear operator. The discussion will therefore be restricted to linear operators.

The factorization problem is one of considerable interest and importance, and the natural setting for its study appears to be a Banach algebra. Thus, assume that the operators under consideration form a Banach algebra. As is easily verified, the causal operators will then form a subalgebra since causal operators are closed under addition, under composition, and under multiplication by scalars. This is the reason for introducing the projection operators and stating the theorem in terms of arbitrary projections and elements of a Banach Algebra.

The general factorization theorem thus obtained is then specialized to certain classes of linear operators in Hilbert space. It is also shown

---

[16] These problems permeate the work of Wiener (Ref. 30) and Krein (Ref. 31).

that in the case of certain convolution operators with a time-invariant kernel the results are rather conservative, and less restrictive factorization theorems due to Krein (Ref. 31) exist. The setting of the factorization problem is the same as used by Zames and Falb (Ref. 14), but the results are more general. The method of proof is inspired by a paper by Baxter (Ref. 32) in probability theory.

Definitions: The relevant notions of Banach algebras have been introduced in Section 2.5. Let σ be a Banach algebra with a unit I. A bounded linear transformation, π, from σ into itself is said to be a *projection* on σ if $\pi^2 = \pi$ and if the range of π forms a subalgebra of σ. Note that the range of a projection is thus assumed to be closed under addition *and* multiplication. The *norm* of π, denoted by $\|\pi\|$, is defined in the usual way as the greatest lower bound of all numbers M which satisfy for all $A \in \sigma$ the inequality $\|\pi A\| \leqslant M \|A\|$.[17] The identity transformation on σ is denoted by θ.

The following factorization theorem states the main result of this section.[18]

THEOREM 3.18

Let σ be a linear Banach algebra with a unit element I and let π^+ and $\pi^- = \theta - \pi^+$ be projections on σ. Let σ^+ and σ^- be the ranges of π^+ and π^-, and assume that $\|\pi^+\| \leqslant 1$ and that $\|\pi^-\| \leqslant 1$. Let Z be an element of σ and let ρ be a scalar. *If* $\|Z\| < |\rho|$, then there exist elements $Z^+ \in \sigma$ and $Z^- \in \sigma$ such that:

1. $M = \rho I - Z = Z^- Z^+$;
2. Z^+ and Z^- are invertible in σ; and
3. Z^+ and $(Z^+)^{-1}$ belong to $\sigma^+ \oplus I$,[19] and Z^- and $(Z^-)^{-1}$ belong to $\sigma^- \oplus I$.

In order to prove Theorem 3.18, Lemmas 3.8–3.10 will be established first.

---

[17] Contrary to Hilbert spaces, projections in Banach spaces need *not* have norm less than or equal to one. Since the spaces under consideration here are spaces of bounded linear operators, there is in general no inner product structure on them, and the condition $\|\pi^+\| \leqslant 1$ thus becomes a constraint.

[18] A similar theorem, due to Masani, with the estimate $\|Z\| < \rho/4$ has appeared in the literature (Ref. 33). Another interesting factorization theorem in a somewhat different setting has been obtained by Gohberg and Krein (Ref. 35).

[19] The notation $\sigma^+ \oplus I$ means all elements of σ which are of the form $R + aI$ with $R \in \sigma^+$ and a a scalar; $\sigma^- \oplus I$ is similarly defined.

LEMMA 3.8

Let $\{A_k\}$, $\{P_k\}$, and $\{N_k\}$, $k = 1, 2, \ldots$ be sequences of elements of σ, σ^+, and σ^- respectively and assume that for some $r_0 > 0$ and all $|r| \leqslant r_0$: 1, The series $A = I + \sum_{k=1}^{\infty} A_k r^k$, $P = I + \sum_{k=1}^{\infty} P_k r^k$, and $N = I + \sum_{k=1}^{\infty} N_k r^k$ converge; and 2, the elements are related by $A = PN$. Then A uniquely determines the sequences $\{P_k\}$ and $\{N_k\}$.

Proof: Equating coefficients of equal powers in r in the equality $A = PN$ leads to $P_1 + N_1 = A_1$ and $P_n + N_n = A_n - \sum_{k=1}^{n-1} P_k N_{n-k}$ for $n = 2, 3, \ldots$. Thus $P_n = \pi^+(A_n - \sum_{k=1}^{n-1} P_k N_{n-k})$ and $N_n = \pi^-(A_n - \sum_{k=1}^{n-1} P_k N_{n-k})$ which shows that A uniquely determines P_n and N_n provided it uniquely determines $P_1, \ldots, P_{n-1}$ and $N_1, \ldots, N_{n-1}$. Since A uniquely determines P_1 and N_1 by $P_1 = \pi^+ A_1$ and $N_1 = \pi^- A_1$, the result follows by induction.

LEMMA 3.9

The equations $P = I + r\pi^+(ZP)$ and $N = I + r\pi^-(NZ)$ have a unique solution $P \in \sigma$ and $N \in \sigma$ for all $|r| \leqslant |\rho|^{-1}$. Moreover, these solutions are given by the convergent series $P = \sum_{k=1}^{\infty} P_k r^k$ and $N = \sum_{k=0}^{\infty} N_k r^k$ with $P_0 = N_0 = I$, $P_{k+1} = \pi^+(ZP_k)$, and $N_{k+1} = \pi^-(N_k Z)$. Also, $P \in \sigma^+ \oplus I$ and $N \in \sigma^- \oplus I$.

Proof: The result follows from the inequalities

$$\|r\pi^+(Z(A - B))\| \leqslant |\rho|^{-1} \|Z\| \|A - B\|,$$

$$\|r\pi^-((A - B)Z)\| \leqslant |\rho|^{-1} \|Z\| \|A - B\|,$$

and the contraction mapping principle. Moreover, it is easily verified that the successive approximations induced by this contraction mapping with $P_0 = N_0 = I$ yield the power series expressions of P and N as claimed in the lemma.

LEMMA 3.10

The solutions P and N to the equations of Lemma 3.9 are invertible for all $|r| \leqslant |\rho|^{-1}$, the inverses being given by $P^{-1} = I - r\pi^+(NZ)$ and $N^{-1} = I - r\pi^-(ZP)$. Moreover, $N^{-1}P^{-1} = I - rZ$ for all $|r| \leqslant |\rho|^{-1}$. Also, $P^{-1} \in \sigma^+ \oplus I$ and $N^{-1} \in \sigma^- \oplus I$.

Proof: From the equations defining P and N it follows for $|r| \leqslant |\rho|^{-1}$ that

$$\|r\pi^+(NZ)\| \leqslant \frac{|r| \|Z\|}{1 - |r| \|Z\|} \quad \text{and} \quad \|r\pi^-(ZP)\| \leqslant \frac{|r| \|Z\|}{1 - |r| \|Z\|} .$$

Since all elements of σ which are of the form $I - B$ with $\|B\| < 1$ are invertible, it thus follows that $I - r\pi^+(NZ)$, $I - r\pi^-(ZP)$ and $I - rZ$ are invertible for $|r| \leqslant |\rho|^{-1}/2$. Furthermore, the inverses are given by the convergent series

$$(I - r\pi^+(NZ))^{-1} = I + \sum_{k=1}^{\infty} (\pi^+(NZ))^k r^k,$$

$$(I - r\pi^-(ZP))^{-1} = I + \sum_{k=1}^{\infty} (\pi^-(ZP))^k r^k$$

$$(I - rZ)^{-1} = I + \sum_{k=1}^{\infty} Z^k r^k$$

From the equations of P and N, it follows that $(I - rZ)P = I - r\pi^-(ZP)$ and $N(I - rZ) = I - r\pi^+(NZ)$ for $|r| \leqslant |\rho|^{-1}$ and thus that $(I - rZ)^{-1} = P(I - r\pi^-(ZP))^{-1} = (I - r\pi^+(NZ))^{-1}N$ for $|r| < |\rho|^{-1}/2$. Since all factors in the equalities are given by the convergent series given in this lemma and in Lemma 3.9 and since σ^+ and σ^- are closed under multiplication, Lemma 3.8 is applicable. This yields $P = (I - r\pi^+(NZ))^{-1}$, $N = (I - r\pi^-(ZP))^{-1}$ and $PN = (I - rZ)^{-1}$ for $|r| \leqslant |\rho|^{-1}/2$. Thus for $|r| \leqslant |\rho|^{-1}/2$ the following equalities hold:

$$P(I - r\pi^+(NZ)) = (I - r\pi^+(NZ))P = I,$$

$$N(I - r\pi^-(ZP)) = (I - r\pi^-(ZP))N = I,$$

$$(I - r\pi^-(ZP))(I - r\pi^+(NZ)) = I - rZ.$$

Since, for $|r| \leqslant |\rho|^{-1}$, all terms in the above equalities are given by geometrically convergent power series in r, they are analytic functions of r for $|r| \leqslant |\rho|^{-1}$. Since equality holds for $|r| \leqslant |\rho|^{-1}/2$ it is thus concluded from analyticity that equality holds for all $|r| \leqslant |\rho|^{-1}$.

PROOF OF THEOREM 3.18

Let $r = \rho^{-1}$ in the Lemma 3.10. The theorem follows with $Z^- = \rho(I - \rho^{-1}\pi^-(ZP))$, $(Z^-)^{-1} = \rho^{-1}N$, $Z^+ = I - \rho^{-1}\pi^+(NZ)$ and $(Z^+)^{-1} = P$.

Under a suitable choice of the Banach algebra and the projection operators a number of interesting corollaries to Theorem 3.18 hold, two of which will now be stated.

Let R be an element of $\mathscr{L}(l_2, l_2)$ and let $\{r_{kl}\}$, k, $l \in I$ be the corresponding array. R is said to belong to $\mathscr{L}^+(l_2, l_2)$ if $r_{kl} = 0$ for all $k < l$. It is said to belong to $\mathscr{L}^-(l_2, l_2)$ if R^* belongs to $\mathscr{L}^+(l_2, l_2)$. Elements of

$\mathscr{L}^+$ correspond to causal elements of $\mathscr{L}$ whereas elements of $\mathscr{L}^-$ correspond to anticausal elements of $\mathscr{L}$.

COROLLARY 3.18.1

Let Z be an element of $\mathscr{L}(l_2,l_2)$ such that $Z - \epsilon I$ is doubly dominant for some $\epsilon > 0$. Then there exist elements M and N of $\mathscr{L}(l_2,l_2)$ such that:

1. $Z = MN$;
2. M and N have bounded inverses M^{-1} and N^{-1}; and
3. N and N^{-1} belong to $\mathscr{L}^+(l_2,l_2)$, and M and M^{-1} belong to $\mathscr{L}^-(l_2,l_2)$.[20]

COROLLARY 3.18.2

Let $A(z) - \epsilon$ be the z-transform of a sequence which is dominant for some $\epsilon > 0$. Then there exist functions $A^+(z)$ and $A^-(z)$ such that

1. $A(z) = A^-(z)A^+(z)$; and
2. $A^+(z)$ and $(A^+(z))^{-1}$ are the z-transforms of l_1-sequences $\{a_k^+\}$ and $\{b_k^+\}$ with $a_k^+ = b_k^+ = 0$ for $k < 0$, and $A^-(z)$ and $(A^-(z))^{-1}$ are the z-transforms of l_1-sequences $\{a_k^-\}$ and $\{b_k^-\}$ with $a_k^- = b_k^- = 0$ for $k > 0$.

Proof: Both of these corollaries follow from Theorem 3.18 under a suitable choice of the Banach algebra σ and the projections π^+ and π^-.

Corollary 3.18.1 follows from Theorem 3.18 with the Banach algebra σ all members of $\mathscr{L}(l_2,l_2)$ such that if $A \in \sigma$ and if $\{a_{kl}\}$, k, $l \in I$ is the corresponding array, then the sequences $\{a_{kl}\}$ belong to l_1 for fixed k and l, uniformly in k and l; i.e., there exists an M such that $\sum_{k=-\infty}^{+\infty} |a_{kl}| \leqslant M$ and $\sum_{l=-\infty}^{+\infty} |a_{kl}| \leqslant M$. Multiplication is defined in the usual way as composition of elements of $\mathscr{L}(l_2,l_2)$. The norm is

---

[20] It comes somewhat as a surprise that Theorem 3.18 when applied to linear operators on $\mathscr{L}_2$ yields factorization of doubly infinite matrices, just sufficient to ensure factorizability of the positive operators discovered earlier in this chapter. This result also hinges on the somewhat unexpected type of norm chosen in applying Theorem 3.18. Notice that this norm is *not* the one induced by the l_2-topology. The reason for not using the induced norm is that it was not possible to show that $\|\pi^+\| \leqslant 1$ in this topology. Whether or not this is true remains unclear. For time invariant operators this question reduces to proving or disproving that for any $g \in L_1$,

$$\max_{\omega \in R} \left| \int_{-\infty}^{+\infty} g(t)e^{-j\omega t}\,dt \right| \geqslant \max_{\omega \in R} \left| \int_0^\infty g(t)e^{-j\omega t}\,dt \right|.$$

If this inequality holds then Theorem 3.18 can be used to show that any strictly positive operator on a Hilbert space admits a factorization into a causal and an anticausal part.

defined as the greatest lower bound of all numbers M satisfying the above inequalities. The nonobvious elements in verifying that σ forms a Banach algebra are that σ is closed under multiplication, that $\|AB\| \leqslant \|A\| \|B\|$ for all $A, B \in \sigma$, and that σ is complete. Closedness under multiplication follows from Fubini's Theorem for sequences (Ref. 22, p. 245) and the inequalities

$$\sum_{k=-\infty}^{+\infty} \left| \sum_{i=-\infty}^{+\infty} a_{ki} b_{il} \right| \leqslant \sum_{k=-\infty}^{+\infty} \sum_{i=-\infty}^{+\infty} |a_{ki}| |b_{il}|$$

$$= \sum_{i=-\infty}^{+\infty} |b_{il}| \sum_{k=-\infty}^{+\infty} |a_{ki}|$$

$$\leqslant \|A\| \|B\|$$

and

$$\sum_{l=-\infty}^{+\infty} \left| \sum_{i=-\infty}^{+\infty} a_{ki} b_{il} \right| \leqslant \|A\| \|B\|.$$

These inequalities also show that $\|AB\| \leqslant \|A\| \|B\|$. Completeness follows from the fact that l_1 is complete. The projection operator π^+ is defined by $\pi^+ A = B$; if $\{a_{kl}\}$ and $\{b_{kl}\}$, $k, l \in I$ are the corresponding arrays, then $a_{kl} = b_{kl}$ for all $k \geqslant l$, and $b_{kl} = 0$ otherwise. The operator π^- is defined by $\pi^- = \theta - \pi^+$. It is clear that $\|\pi^+\| = 1$ and that $\|\pi^-\| = 1$. The only fact that is left to be shown is that if $Z - \epsilon I$ is doubly dominant for some $\epsilon > 0$, then Z can be written as $Z = \rho I - A$ with $\|A\| < \rho$. It is easily verified that any ρ with $|\rho| \geqslant \sup_{k \in I} z_{kk}$ yields such a decomposition.

The proof of Corollary 3.18.2 is completely along the lines of the proof of Corollary 3.18.1 with the Banach algebra σ consisting of all l_1 sequences, and with multiplication of $A = \{a_k\}$ and $B = \{b_k\}$ defined by $AB = C = \{c_k\}$, where $c_k = \sum_{l=-\infty}^{+\infty} a_{k-l} b_l$ and $\|A\| = \sum_{k=-\infty}^{+\infty} |a_k|$. The projection operator π^+ is defined by $\pi^+ A = B$ with $A = \{a_k\}$, $B = \{b_k\}$, $b_k = a_k$ for $k \geqslant 0$, and $b_k = 0$ for $k < 0$. The operator π^- is defined by $\pi^- = \theta - \pi^+$.

The factorization in Corollary 3.18.2 is valid under much weaker conditions than stated. Indeed, although dominance of the involved sequence is certainly sufficient for the factorization to be possible, it is by no means necessary — as is shown by the following theorem, due to Krein (Ref. 31).

THEOREM 3.19

Let $A(z)$ be the z-transform of an l_1-sequence. Then there exist functions $A^+(z)$ and $A^-(z)$ such that:

1. $A(z) = A^-(z)A^+(z)$, and
2. $A^+(z)$ and $(A^+(z))^{-1}$ are the z-transforms of l_1-sequences $\{a_k^+\}$ and $\{b_k^+\}$ with $a_k^+ = b_k^+ = 0$ for $k < 0$, and $A^-(z)$ and $(A^-(z))^{-1}$ are the z-transforms of l_1-sequences $\{a_k^-\}$ and $\{b_k^-\}$ with $a_k^- = b_k^- = 0$ for $k > 0$,

if and only if $A(z) \neq 0$ for $|z| = 1$ and the increase in the argument of the function $A(z)$ as z moves around the circle $|z| = 1$ is zero. Moreover, all factorizations which satisfy conditions 1 and 2 differ only by a nonzero multiplicative constant.

The factorization analogous to those obtained in Corollaries 3.18.1 and 3.18.2 for operators on L_2 with time-varying kernels is straightforward and will not be explicitly given. The analogue to Theorem 3.19 for operators with a time-invariant kernel is Theorem 3.20; Theorem 3.20, which is also due to Krein, is less restrictive than the analogous factorization obtained in Theorem 3.18.

Let $\mathscr{G}_1$ be a class of operators from L_2 into itself each element of which is determined by an element $(g(t),\{g_k\})$ of $L_1 \times l_1$ and by a mapping $\{t_k\}$ from I into R. The operator $G \in \mathscr{G}_1$ maps $x \in L_2$ into y, with $y(t) = \sum_{k=-\infty}^{+\infty} g_k x(t - t_k) + \int_{-\infty}^{+\infty} g(t - \tau)x(\tau)\,d\tau$. It is simple to verify that G is well defined, i.e., that it maps L_2 into itself. Let $\mathscr{G}_1^+$ denote the subclass of $\mathscr{G}_1$ for which the determining element of $L_1 \times l_1$ and the mapping $\{t_k\}$ satisfy $g(t) = 0$ for $t < 0$ and $t_k \geqslant 0$ for all $k \in I$. Let $\mathscr{G}_1^-$ denote the subclass of $\mathscr{G}_1$ for which the determining element of $L_1 \times l_1$ and the mapping $\{t_k\}$ satisfy instead $g(t) = 0$ for $t > 0$ and $t_k \leqslant 0$ for all $k \in I$. Clearly $G \in \mathscr{G}_1^+$ if and only if $G^* \in \mathscr{G}_1^-$.

THEOREM 3.20

Let $G \in \mathscr{G}_1$ and assume that all the delays are equally spaced, i.e., that $t_k = kT$ for some k. Then there exist elements $G^+ \in \mathscr{G}_1$ and $G^- \in \mathscr{G}_1$ such that:

1. $G = G^-G^+$,
2. G^+ and G^- are invertible, and
3. G^+ and $(G^+)^{-1} \in \mathscr{G}_1^+$, and G^- and $(G^-)^{-1} \in \mathscr{G}_1^-$,

if and only if, $|G(j\omega)| \geqslant \epsilon$, for some $\epsilon > 0$ and all $\omega \in R$, and the increase in the argument of the function $G(j\omega)$ as ω varies from $-\infty$ to $+\infty$ is zero.[21]

A slightly weaker version of Theorem 3.20 is given by Krein (Ref. 31). However, the extension of his proof to cover the form given in Theorem 3.20 presents no apparent difficulties.

The main interest in the above factorization of operators stems from the fact that it leads to the solution of the so-called Wiener-Hopf equation. The Wiener-Hopf equation plays a central role in applied mathematics (Ref. 34), and its importance in the analytical design of engineering systems cannot be overestimated. It is the logical conclusion of the optimization of quadratic functionals subject to linear constraints. In fact, the celebrated Riccati equation — which plays a central role in control and detection theory — can be viewed as a method of solving a Wiener-Hopf equation. Among its applications are the optimal control of linear systems under quadratic performance criteria, minimum variance filtering of Gaussian processes, network synthesis, solution of boundary-value problems, etc.

In the abstract notation of this section, application of the Wiener-Hopf equation amounts to asking for the solution X of the system of equations

$$\pi^+(XR) = \pi^+M \qquad \text{and} \qquad \pi^+X = X.$$

Theorem 3.21 below shows how this solution can be readily obtained, provided that R admits a suitable factorization. First, however, a simple lemma will be proven:

LEMMA 3.11

Let σ be a linear algebra with a unit element I and let π^+ and $\pi^- = \theta - \pi^+$ be projections on σ with ranges σ^+ and σ^- respectively. Let $A \in \sigma$ and $B \in \sigma^- \oplus I$. Then $\pi^+((\pi^+A)B) = \pi^+(AB)$.

[21] Let $G(j\omega)$, $\omega \in R$, be a complex-valued function of ω with $G(j\omega) = A(j\omega) + L(j\omega)$ and $A(j\omega)$ periodic and $\lim_{\omega \to \pm\infty} L(j\omega) = 0$. The increase in the argument of the function $G(j\omega)$ is said to be zero if

$$\lim_{N \to \infty} \int_{2\pi N/T}^{2\pi N/T} d \arg G(j\omega) = \lim_{\omega \to \infty} \frac{1}{2\omega} \int_{-\omega}^{\omega} d \arg A(j\omega) = 0.$$

This definition is a natural generalization of the usual definition of the increase in the argument of a complex-valued function.

Proof: Since

$$\pi^+(AB) = \pi^+((\pi^+A + \pi^-A)B)$$
$$= \pi^+((\pi^+A)B) + \pi^+((\pi^-A)B) = \pi^+((\pi^+A)B),$$

the lemma follows.

THEOREM 3.21

Let σ be a linear algebra with a unit element I and let π^+ and $\pi^- = \theta - \pi^+$ be projections on σ with ranges σ^+ and σ^-, respectively. Let R and M be given elements of σ, and assume that there exist elements $R^+ \in \sigma$ and $R^- \in \sigma$ such that:

1. $R = R^+R^-$;
2. R^+ and R^- are invertible in σ; and
3. R^+ and $(R^+)^{-1}$ belong to $\sigma^+ \oplus I$, and R^- and $(R^-)^{-1}$ belong to $\sigma^- \oplus I$.

Then the unique solution of the Wiener-Hopf equation, i.e., the solution of the system of equations $\pi^+(XR) = \pi^+M$, $\pi^+X = X$, is given in terms of this factorization of R by $X = (\pi^+(M(R^-)^{-1}))(R^+)^{-1}$.

Proof: It will first show that $X = (\pi^+(M(R^-)^{-1}))(R^+)^{-1}$ is a solution of the Wiener-Hopf equation. Clearly $\pi^+X = X$. Moreover,

$$\pi^+(XR) = \pi^+((\pi^+(M(R^-)^{-1}))(R^+)^{-1}R)$$
$$= \pi^+((\pi^+(M(R^-)^{-1}))R^-) = \pi^+((M(R^-)^{-1})R^-) = \pi^+M.$$

These equalities follow from Lemma 3.11 since $(R^+)^{-1} \in \sigma^+ \oplus I$ and $R^- \in \sigma^- \oplus I$. To show that $(\pi^+(M(R^-)^{-1}))(R^+)^{-1}$ is the unique solution of the Wiener-Hopf equation, assume that $\pi^+(XR) = \pi^+M$ and $\pi^+X = X$. Then $(\pi^+(\pi^+(XR^+R^-)))(R^-)^{-1} = \pi^+(M(R^-)^{-1})$ since $(R^-)^{-1} \in \sigma^- \oplus I$. Hence, since $R^+ \in \sigma^+ \oplus I$, the equalities $\pi^+(M(R^-)^{-1}) = \pi^+(XR^+) = XR^+$ hold, which shows that $(\pi^+(M(R^-)^{-1}))(R^+)^{-1}$ is indeed the unique solution of the Wiener-Hopf equation.

3.8 Positive Operators on Extended Spaces

As defined in Section 2.5, a causal operator F from L_{2e} into itself is said to be (incrementally) positive on L_{2e} if for all $T \in S$, $P_T F P_T$ is (incrementally) positive on L_2. As was pointed out there, the following relation between positivity on L_2 and L_{2e} exists: a causal operator on L_{2e} which in addition maps L_2 into itself is (incrementally) positive on L_{2e} if and only if it is (incrementally) positive on L_2. Hence nothing new is required to examine the positivity of operators on L_{2e} *provided* they map L_2 into itself.

For causal operators on L_{2e} which do *not* map L_2 into itself it is, however, a simple matter to generalize the theorems of the previous situation to operators on L_{2e}.[22] For example, Theorem 3.2 can be extended to cover the case where $K(t) \in L_{\infty e}$ (it suffices for $K(t)$ to belong to $L_{\infty e}$ in order for the operator $(Kx)(t) = K(t)x(t)$ to be well defined as a map from L_{2e} into itself). The analogue of Theorem 3.11 is also obvious and requires the analogue of the dominance conditions *without* the requirement that $\sup_{k \in I} m_{kk} < \infty$; instead, it is assumed that $m_{kl} = 0$ for $k < l$. This analogue, however, does not fully exploit the power of the rearrangement inequalities derived in this chapter, and it is in fact possible to prove a somewhat stronger positive-operator theorem, which does not require boundedness of the operators involved.

Definitions: Let $M = \{m_{kl}\}$, with $k, l \in I^+$, be a square array of real numbers. Then M is said to be *doubly hyperdominant* if $m_{kl} \leqslant 0$ for $k \neq l$ and if $\sum_{k=0}^{\infty} m_{kl}$ and $\sum_{l=0}^{\infty} m_{kl}$ exist and are nonnegative for all l and k. M is said to be *doubly dominant* if $m_{ll} \geqslant \sum_{k=0, k \neq l}^{\infty} |m_{kl}|$ and $m_{kk} \geqslant \sum_{k=0, l \neq k}^{\infty} |m_{kl}|$. Every doubly dominant array M, with $k, l \in I^+$ and $m_{kl} = 0$ for $k < l$, defines a causal linear operator from $l_{2e}(0, \infty)$ into itself with $(Mx)_k = \sum_{l=0}^{k} m_{kl} x_l$. This operator need not, however, be bounded on l_{2e}. Given two square arrays of real numbers $M_1 = \{m_{kl}^{(1)}\}$ and $M_2 = \{m_{kl}^{(2)}\}$, $k, l \in I^+$, it is possible to define the product $M_1 M_2$ as the square array with $(M_1 M_2)_{kl} \triangleq \sum_{i=0}^{\infty} m_{ki}^{(1)} m_{il}^{(2)}$ whenever this *infinite sum exists* for all k and l, and thus in particular whenever M_1 is causal (i.e., whenever M_1 defines a causal operator on l_{2e}; in other words, whenever $m_{kl}^{(1)} = 0$ for $k < l$), since the summation defining this product then becomes finite. Let $M = \{m_{kl}\}$, $k, l \in I^+$, be given; then $M^* = \{n_{kl}\}$, $k, l \in I^+$ is defined as the square array with $n_{kl} = m_{lk}$ for all k and l.

THEOREM 3.22

Let L_1 and L_2 be causal linear operators from $l_{2e}(0, \infty)$ into itself and let Γ denote the operator on $l_{2e}(0, \infty)$ defined by $(Fx)_k = f(x_k)$ with f a map from R into itself. Assume that L_1 has a causal inverse on $l_{2e}(0, \infty)$.

---

[22] It should be pointed out that Theorem 3.1 is an exception to this, i.e., that, in a sense, the *bounded* positive time-invariant causal convolution operators on L_{2e} define in fact the most general positive time-invariant causal operators on L_{2e} as well. More specifically, consider the convolution operator (as in Theorem 3.1) with vanishing kernel for $t < 0$ (for causality) and whose kernel is locally integrable (i.e., integrable on compact sets) so that it defines a causal linear time-invariant operator on L_{2e}. If for some $\Delta T > 0$, the integral of the norm of the kernel on the interval $[T, T + \Delta T]$ becomes unbounded as $T \to \infty$, then this operator will *not* be positive on L_{2e}.

Then L_1FL_2 is a positive operator on $l_{2e}(0,\infty)$ for any monotone nondecreasing (odd-monotone nondecreasing) first and third quadrant function f if and only if the square array determined by $L_2(L_1^{-1})^*$ is doubly hyperdominant (dominant).

Proof: Let N be a nonnegative integer, let $x \in l_{2e}(0,\infty)$, and let $P_N x$ denote the truncation of x. The operator $P_N L_1FL_2 P_N$ is positive on R^N if and only if $P_N L_1^{-1} P_N L_1 FL_2 P_N (L_1^{-1})^* P_N$ is. Since, however,

$$P_N L_1^{-1} P_N L_1 FL_2 P_N (L_1^{-1})^* P_N = P_N FL_2 (L_1^{-1})^* P_N,$$

the theorem follows from Theorem 3.10.

References

1. Canales, R., *A Priori Bounds on the Performance of Optimal Systems*, Technical Memorandum No. ESL-TM-362, Electronic Systems Laboratory, Massachusetts Institute of Technology, Cambridge, Mass, 1968.
2. Titchmarsh, E. C., *Introduction to the Theory of Fourier Integrals*, Oxford University Press, Oxford, 1962.
3. Gruber, M. and Willems, J. L., "On a Generalization of the Circle Criterion," *Proceedings of the Fourth Annual Allerton Conference on Circuit and System Theory*, pp. 827–835, 1966.
4. Freedman, M., and Zames, G., "Logarithmic Variation Criteria for the Stability of Systems with Time-Varying Gains," *SIAM J. on Control*, Vol. 6, pp. 487–507, 1968.
5. Page, C. H., "Frequency Conversion with Positive Nonlinear Resistors," *Journal of Research of the National Bureau of Standards*, Vol. 56, No. 4, pp. 179–182, 1956.
6. Pantell, R. H., "General Power Relationships for Positive and Negative Nonlinear Resistive Elements," *Proceedings IRE*, Vol. 46, pp. 1910–1913, 1958.
7. Black, W. L., "Some New Power Frequency Inequalities for Nonlinear Capacitive Harmonic Multipliers," *IEEE Proceedings*, Vol. 54, pp. 1995–1996, 1966.
8. Brockett, R. W., and Willems, J. L., "Frequency Domain Stability Criteria: Part I and II," *IEEE Trans. on Automatic Control*, Vol. AC-10, pp. 255–261 and 401–413, 1965.
9. Narendra, K. S., and Neuman, C. P., "Stability of a Class of Differential Equations with a Single Monotone Nonlinearity," *SIAM J. on Control*, Vol. 4, pp. 295–308, 1966.
10. Zames, G., "On the Input-Output Stability of Time-Varying Nonlinear Feedback Systems. Part I: Conditions Derived Using Concepts of Loop Gain, Conicity, and Positivity. Part II: Conditions Involving Circles in the Frequency Plane and Sector Nonlinearities," *IEEE Trans. on Automatic Control*, Vol. AC-11, pp. 228–238 and 465–476, 1966.

11. O'Shea, R. P., "A Combined Frequency-Time Domain Stability Criterion for Autonomous Continuous Systems," *IEEE Trans. on Automatic Control*, Vol. AC-11, pp. 477–484, 1966.
12. O'Shea, R. P., "An Improved Frequency Time Domain Stability Criterion for Autonomous Continuous Systems," *IEEE Trans. on Automatic Control*, Vol., AC-12, pp. 725–731, 1967.
13. O'Shea, R. P., and Younis, M. I., "A Frequency Time Domain Stability Criterion for Sampled-Data Systems," *IEEE Trans. on Automatic Control*, Vol. AC-12, pp. 719–724, 1967.
14. Zames, G., and Falb, P. L., "Stability Conditions for Systems with Monotone and Slope-Restricted Nonlinearities," *SIAM J. on Control*, Vol. 6, pp. 89–108, 1968.
15. Hardy, G. H., Littlewood, T. E., and Polya, G., *Inequalities* (second edition), Cambridge University Press, Cambridge, England, 1952.
16. Prosser, R. T., *An Inequality for Certain Correlation Functions*, M.I.T. Lincoln Laboratory Group Report No. 64-63, Lexington, Mass., 1964.
17. Edelmann, H., "Ein Satz über Stieltjes-Matrizen und seine Netzwerktheoretische Deutung," *Archiv für Elektrische Uebertragung* 23-1, pp. 59–60, 1969.
18. Varga, G. S., *Matrix Iterative Analysis*, Prentice-Hall, Englewood Cliffs, N.J., 1962.
19. Bellman, R., *Introduction to Matrix Analysis*, McGraw-Hill, New York, 1960.
20. Marcus, M., and Minc, H., *A Survey of Matrix Theory and Matrix Inequalities*, Allyn and Bacon, Boston, 1964.
21. Akhiezer, N. I., and Glazman, I. M., *Theory of Linear Operators in Hilbert Space*, Frederick Ungar, New York, 1961.
22. Cooke, R. G., *Infinite Matrices and Sequence Spaces*, Dover Publications, New York, 1955.
23. Grenander, U., and Szegö, G., *Toeplitz Forms and Their Applications*, University of California Press, Berkeley, 1958.
24. Beckenbach, E., and Bellman, R., *Inequalities*, Springer-Verlag, Berlin, 1961.
25. Willems, J. C., and Gruber, M., "Comments on a Combined Frequency-Time Domain Stability Criterion for Autonomous Continuous Systems," *IEEE Trans. on Automatic Control*, Vol. AC-12, pp. 217–218, 1967.
26. Feller, W., *An Introduction to Probability Theory and Its Applications*, Vol. II, John Wiley and Sons, New York, 1966.
27. Penfield, P., Jr., *Frequency-Power Formulas*, The M.I.T. Press, Cambridge, Mass., 1960.
28. Manley, J. M., and Rowe, H. E., "Some General Properties of Nonlinear Elements, I. General Energy Relations," *Proceedings IRE*, Vol. 44, pp. 904–913, 1956.
29. Carroll, J. E., "A Simplified Derivation of the Manley and Rowe Power Relationships," *J. Elect. and Control*, Vol. 6, No. 4, pp. 359–361, 1959.
30. Wiener, N., *Time Series*, The M.I.T. Press, Cambridge, Mass., 1964.
31. Krein, M. G., "Integral Equations on a Half-Line with Kernels Depending upon the Difference of the Arguments," *American Mathematical Society Translations*, Series 2, Vol. 22, pp. 163–288, 1962.

32. Baxter, G., "An Operator Identity," *Pacific J. of Mathematics*, Vol. 8, pp. 649–663, 1958.
33. Masani, P., "The Laurent Factorization of Operator-Valued Functions," *Proc. London Math. Soc.*, Vol. 3, No. 6, pp. 59–69, 1956.
34. Noble, B., *Methods Based on the Wiener-Hopf Technique*, Pergamon Press, New York, 1958.
35. Gohberg, I. C., and Krein, M. G., "On the Factorization of Operators in Hilbert Space," *American Mathematical Society Translations*, Series 2, Vol. 51, pp. 155–188, 1966.
36. Hille, E., and Phillips, R. S., *Functional Analysis and Semi-Groups* (second edition), American Mathematical Society, Providence, 1957.

4 Feedback Systems

4.1 Introduction

Feedback remains the basic concept of control: some output variables of the system to be controlled are measured, and this information is processed to generate an input to the system to be controlled such that the overall system behaves in some desired fashion. Feedback control thus involves cybernetics: observation, information processing, decision, and execution. It is believed that Norbert Wiener first had the insight to conclude that such cybernetic situations are intimately related to uncertainty: in the absence of uncertainty, there would be no need for feedback, and all decisions could be made "open loop."

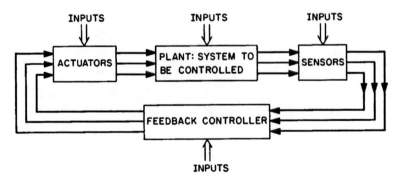

Figure 4.1   Control Loop

However, no system (plant, sensor, or actuator) is free of uncertainty, and control engineers always resort, in the end, to feedback controllers.

In this chapter the basic properties of feedback systems are studied: these include their well-posedness, stability or instability, and continuity or discontinuity. Also, sensitivity is briefly mentioned. The feedback system to be studied is shown in Figure 4.1.

4.2 Mathematical Framework

A model applicable to most feedback systems is shown in Figure 4.2.[1] The functional equations describing this feedback system are

$$e_1 = u_1 - y_2,$$

$$e_2 = u_2 + y_1, \tag{FE}$$

$$y_1 = G_1 e_1, \text{ and}$$

$$y_2 = G_2 e_2,$$

where u_1, u_2 are called the *inputs;* e_1, e_2 are called the *errors*, and y_1, y_2 are called the *outputs*.

Let T_0 be a (finite) real number and let S, the *time-interval of definition*, be $[T_0, \infty)$.[2] Let B_1 and B_2 be given Banach spaces, and let $Y(B_i)$,

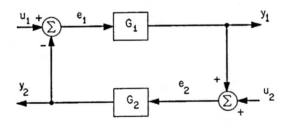

Figure 4.2   The Feedback System under Consideration

[1] The assumptions implied by this configuration are: 1, additivity of the inputs and 2, only two input summing junctions are present in the loop. The methods presented here lend themselves easily to the study of more general configurations. The philosophy of using this configuration is that one assumes additive driving noise and additive noise in the measurements, the usual situation studied, for example, in combined estimation and control problems.

[2] The extension to discrete systems is a trivial one. The case $S = (-\infty, +\infty)$ is not treated since the author is not convinced that this case is of physical significance. Some remarks regarding this case will be made in Section 4.6.

$i = 1, 2$, denote the linear space of B_i-valued functions defined on S, i.e., $Y(B_i) \triangleq \{x \mid x : S \to B_i\}$.

Definition: Let $T \in S$. Then P_T denotes the projection operator on $Y(B_i)$, $i = 1, 2$, defined for $x \in Y(B_i)$ by

$$(P_T x)(t) \triangleq \begin{cases} x(t) & \text{for } t < T, \, t \in S \\ 0 & \text{otherwise} \end{cases} ;$$

P_T will be called the *truncation operator* and $P_T x$ will be called the *truncation* of x at time T. The collection $\{P_T\}$, $T \in S$, thus consists of a family of projection operators on $Y(B_i)$.

Definition: Let $W_i \subset Y(B_i)$, $i = 1, 2$, be a Banach space. The *extended space,* W_{ie}, is defined as

$$W_{ie} \triangleq \{x \in Y(B_i) \mid P_T x \in W_i \text{ for all } T \in S, \, T \text{ finite}\}.$$

Recall the definitions of a causal, strongly causal, and locally Lipschitz-continuous operator: Let F be a mapping from W_{ie}, $i = 1, 2$, into itself. Then F is said to be *causal* on W_{ie} if $P_T F P_T = P_T F$ on W_{ie}. It is said to be *strongly causal* on W_{ie} if F is causal on W_{ie} and if for all $T \in S$, $\epsilon > 0$, and $T' \in S$, $T' \leqslant T$, there exists a real number $\Delta T > 0$ such that $\|P_{T'+\Delta T}(Fx - Fy)\| \leqslant \epsilon \|P_{T+\Delta T}(x - y)\|$ for any $x, y \in W_{ie}$ with $P_{T'} x = P_{T'} y$. It is said to be *locally Lipschitz continuous* on W_{ie} if for all $T \in S$, $P_T F P_T$ is Lipschitz continuous on W_i. These notions are generalized in an obvious way to operators from W_{1e} to W_{2e}, $W_{1e} \times W_{2e}$ to W_{1e}, etc., and will thus be used freely in the latter context as well.

For various technical reasons a number of additional concepts involving causal operators are needed.

Definitions: Let F be a strongly causal operator from W_{ie}, $i = 1, 2$, into itself. Then F is said to be *strongly causal, uniformly with respect to past inputs*, if for all $T \in S$, $\epsilon > 0$, and $T' \in S$, $T' \leqslant T$, there exist real numbers $\Delta T > 0$ and $K < \infty$ such that $\|(P_{T'+\Delta T} - P_{T'})(Fx - Fy)\| \leqslant K \|P_{T'}(x - y)\| + \epsilon \|(P_{T+\Delta T} - P_{T'})(x - y)\|$ for all $x, y \in W_{ie}$. Locally Lipschitz continuous strongly causal linear operators are in fact strongly causal, uniformly with respect to past inputs.

Assumptions on the spaces. The following assumptions are made on the spaces W_i, $i = 1, 2$:

W.1. The spaces W_i are closed under the family of projections $\{P_T\}$, $T \in S$.

W.2. For any $x \in W_i$, the norm $\|P_Tx\|$ is a monotone nondecreasing function of T which satisfies $\lim_{T\downarrow T_0} \|P_Tx\| = 0$ and $\lim_{T\uparrow\infty} \|P_Tx\| = \|x\|$. The family of projection operators $\{P_T\}$, $T \in S$, is thus assumed to be a resolution of the identity.

W.3. If $x \in W_{ie}$, then $x \in W_i$ if and only if $\sup_{T\in S} \|P_Tx\| = \lim_{T\uparrow\infty} \|P_Tx\| < \infty$.

W.4. For any real numbers t_1, t_2 such that $T_0 \leqslant t_1 \leqslant t_2$, the vector space $W_{i(t_1,t_2)} \triangleq \{x \in W_i \mid x(t) = 0 \text{ for } t \notin [t_1,t_2]\}$ is a closed subspace of W_i and is thus itself a Banach space under the norm of W_i.

Assumptions on the Operators. The following assumptions are made on the operators G_i, $i = 1, 2$:

G.1. The operator G_1 maps W_{1e} into W_{2e}, and G_2 maps W_{2e} into W_{1e}.

G.2. The operators G_i are causal on W_i, i.e., $P_TG_iP_T = P_TG_i$ on W_i for all $T \in S$.

G.3. The operators G_i are locally Lipschitz continuous on W_{ie}, i.e.,

$$\sup_{\substack{x,y\in W_i \\ P_Tx \neq P_Ty}} \frac{\|P_T(G_ix - G_iy)\|}{\|P_T(x - y)\|} < \infty \qquad \text{for all} \qquad T \in S.$$

G.4. The operators G_i satisfy $G_i0 = 0$.[3]

Assumption on the Inputs. The following assumption is made on the inputs u_i, $i = 1, 2$:

I.1. The input u_i belongs to W_{ie}.

The important notions regarding the feedback system under consideration are those of solutions, well-posedness, stability, and continuity. Definitions of these terms are now introduced.

Definition of a solution. Let $u_i \in W_{ie}$, $i = 1, 2$, be given. Then the quadruple $\{e_1,e_2,y_1,y_2\}$ is said to be a *solution* of the feedback equations if:

SO.1. The inputs and outputs e_1, y_2 belong to W_{1e}, and e_2, y_1 belong to W_{2e}.

SO.2. The feedback equations (FE) are satisfied by $\{e_1,e_2,y_1,y_2\}$.

---

[3] Assumption G.4 can always be taken to be satisfied by proper redefinition of the operators.

The above definition of a solution is a natural one. The definition of well-posedness is somewhat more delicate.

Definition of well-posedness. The feedback system described by the equations (FE) is said to be *well posed* if:

WP.1. There exists a unique solution for any pair of inputs $u_i \in W_{ie}$, $i = 1, 2$. Let G denote the operator from $W_{1e} \times W_{2e}$ into itself defined for $x = (x_1, x_2) \in W_{1e} \times W_{2e}$ by $Gx \triangleq (G_2 x_2, -G_1 x_1)$. Condition WP.1 thus requires that the operator $I + G$ be one-to-one and onto, i.e., *invertible* on $W_{1e} \times W_{2e}$.

WP.2. The errors and the outputs depend on the inputs in a *non-anticipatory* way; i.e., $P_T e_1$, $P_T e_2$, $P_T y_1$, and $P_T y_2$ depend, for any $T \in S$, on $P_T u_1$ and $P_T u_2$ only. Conditions WP.1 and WP.2 combined thus require the operator $I + G$ to be invertible on $W_{1e} \times W_{2e}$ and $(I + G)^{-1}$ to be *causal* on $W_{1e} \times W_{2e}$.

WP.3. The errors and the outputs depend, on finite intervals, *Lipschitz continuously* on the inputs. Conditions WP.1, WP.2, and WP.3 combined thus require the operator $I + G$ to be invertible on $W_{1e} \times W_{2e}$ and $(I + G)^{-1}$ to be *causal* and *locally Lipschitz continuous* on $W_{1e} \times W_{2e}$.

WP.4. The errors and the outputs are *insensitive to modeling errors* in the following well-defined sense: Consider the functional equations

$$
\begin{aligned}
e_1 &= S_{1\lambda}(u_1, G_{2\lambda} e_2), \\
e_2 &= T_\varepsilon S_{2\lambda}(u_2, G_{1\lambda} e_1),
\end{aligned}
\qquad \text{(PFE)}
$$

which describes the "physical" feedback system shown in Figure 4.3.

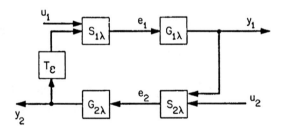

Figure 4.3   The Physical Feedback System

It is assumed that the family of operators, $G_{i\lambda}$, $i = 1, 2$, are parametrized by a parameter λ which is assumed to belong to some index set Λ. The set Λ will be assumed to be a subset of a normed linear space and contain a neighborhood of the origin. Let T_ϵ, $\epsilon < 0$, denote a time delay. The functional equations (PFE) are assumed to approximate the feedback system described by the functional equations (FE) in the sense that for all $\|\lambda\|$ sufficiently small:

1. The operators $S_{1\lambda}$, $S_{2\lambda}$, $G_{1\lambda}$, and $G_{2\lambda}$ map $W_{1e} \times W_{1e}$ into W_{1e}, $W_{2e} \times W_{2e}$ into W_{2e}, W_{1e} into W_{2e}, and W_{2e} into W_{1e}, respectively, and T_ϵ, $\epsilon < 0$, maps W_{1e} into itself by

$$(T_\epsilon x)(t) \triangleq \begin{cases} x(t + \epsilon) & \text{for} \quad t \geqslant T_0 - \epsilon \\ 0 & \text{for} \quad T_0 \leqslant t < T_0 - \epsilon, \end{cases}$$

   i.e., T_ϵ denotes a time delay with delay $-\epsilon > 0$.

2. The operators $S_{i\lambda}$ and $G_{i\lambda}$, $i = 1, 2$, are *locally Lipschitz continuous* on their respective domains.

3. The operators $S_{i\lambda}$ and $G_{i\lambda}$, $i = 1, 2$, are *strongly causal* on their respective domains.

4. The "physical" feedback system described by the equations (PFE) *approximates* the feedback system described by the equation (FE) in the following well-defined sense: For any $T \in S$ and any v_1, $w_1 \in W_{1e}$ and v_2, $w_2 \in W_{2e}$, the operators $S_{i\lambda}$ and $G_{i\lambda}$ satisfy $\lim_{\lambda \to 0} S_{1\lambda}(v_1, w_1) = v_1 - w_1$, $\lim_{\lambda \to 0} S_{2\lambda}(v_2, w_2) = v_2 + w_2$, $\lim_{\lambda \to 0} G_{1\lambda} v_1 = G_1 v_1$, and $\lim_{\lambda \to 0} G_{2\lambda} v_2 = G_2 v_2$.

   It then follows that for any $T \in S$ and $\|\lambda\|$ and $-\epsilon > 0$ sufficiently small:

1. For any input pair $u_i \in W_{ie}$, $i = 1, 2$, there exists a unique solution pair $e_i \in W_{ie}$, $i = 1, 2$, to the equations (PFE). Let $e_i = F_{i\lambda}^\epsilon(u_1, u_2)$, $i = 1, 2$, denote these solutions.

2. The operators $F_{i\lambda}^\epsilon$, $i = 1, 2$, are *causal* on their domain.

3. The operators $F_{i\lambda}^\epsilon$, $i = 1, 2$, are *locally Lipschitz continuous* on their domain.

4. The operators $G_{1\lambda} F_{1\lambda}^\epsilon$ and $T_\epsilon G_{2\lambda} F_{2\lambda}^\epsilon$ are *strongly causal on their domain.*

   It is then required for well-posedness that for any input pair $u_i \in W_{ie}$, $i = 1, 2$, the solutions $F_{i\lambda}^\epsilon$ satisfy

$$\lim_{\epsilon \uparrow 0} \lim_{\lambda \to 0} (F_{1\lambda}^\epsilon(u_1, u_2), F_{2\lambda}^\epsilon(u_1, u_2)) = (I + G)^{-1}(u_1, u_2)$$

Well-posedness is essentially a modeling problem. It expresses that a mathematical model is, at least in principle, adequate as a description of a physical system. The definition which follows is that of stability, which is a *desired* property of a feedback control system.

Definition of stability and instability. The feedback system described by the equations (FE) is said to be *stable* if it satisfies conditions ST.1 and ST.2:

ST.1. It is *well posed.*
ST.2. Any inputs $u_i \in W_i$, $i = 1$, 2, lead to errors and outputs e_1, $y_2 \in W_1$ and e_2, $y_1 \in W_2$.

The system is said to be *stable with finite gain* if it is stable and if in addition:

ST.3. There exist constants ρ_1, $\rho_2 < \infty$ such that for any inputs $u_i \in W_i$, $i = 1, 2$, $\|e_1\|$, $\|e_2\|$, $\|y_1\|$, $\|y_2\| \leqslant \rho_1 \|u_1\| + \rho_2 \|u_2\|$.

The feedback system is said to be *unstable* if:

INST.1. It is *well posed.*
INST.2. It is *not* stable.

A somewhat stronger desired property of a feedback controller is that of continuity.

Definition of continuity and discontinuity. The feedback system described by the equations (FE) is said to be *continuous* if:

C.1. It is *stable.*
C.2. Let $u_i \in W_i$, $i = 1$, 2, be given (but arbitrary), and let e_i and y_i be the corresponding errors and outputs. Then there exists, for any given $\epsilon > 0$, a $\delta > 0$ such that the inputs $u_i + \Delta u_i$ with $\Delta u_i \in W_i$ and $\|\Delta u_1\|$, $\|\Delta u_2\| \leqslant \epsilon$ yield corresponding errors, $e_i + \Delta e_i$, and outputs, $y_i + \Delta y_i$, with Δe_1, $\Delta y_2 \in W_1$, Δe_2, $\Delta y_1 \in W_2$, and $\|\Delta e_1\|$, $\|\Delta e_2\|$, $\|\Delta y_1\|$, $\|\Delta y_2\| \leqslant \delta$.

The system is said to be *Lipschitz continuous* if it is continuous and if in addition:

C.3. There exists a $K < \infty$ (independent of u_i, $i = 1$, 2), such that $\delta = K\epsilon$ will satisfy condition C.2.

The system is said to be *discontinuous* if:

DC.1. It is *well posed.*
DC.2. It is *not* continuous.

4.3 Well-Posedness of Feedback Systems

The question of well-posedness of mathematical models for physical systems is of fundamental importance in engineering.[4]

4.3.1 Justification of the Definition of Well-Posedness

A mathematical model is generally said to be well posed if solutions exist and if these solutions are unique, continuous with respect to the input variables, and continuous with respect to modeling errors (i.e., errors in the parameters of the system). Well-posedness thus imposes a regularity condition on feasible mathematical models for physical systems. Well-posedness expresses the fact that a mathematical model can be, at least in principle, an accurate description of a physical system. In other words, since exact mathematical models would always be well posed, one thus requires this property to be preserved in the modeling.

The definition of well-posedness as given in Section 4.2 requires justification. The assumptions on G imply that it itself be well posed, i.e., that it is a well-defined map, causal, locally Lipschitz-continuous, and a suitable approximation of a physical system. A "physical system" is defined as a locally Lipschitz-continuous strongly causal system. This concept of "physical system" is not standard and calls for an explanation. To require that a physical system be a continuous, well-defined and causal map is natural. The strong causality condition is inspired by the fact that any system exhibits some delay between the input and the output. No system can instantaneously transmit a signal, hence all physical systems have at least an infinitesimal delay. The second point to make about the definition of well-posedness is the precise mathematical interpretation of the idea of a "suitable approximation." For F to approximate F_λ it is required that the operator $F - F_\lambda$ should be small in some appropriate sense. One possible choice — the first one which comes to mind, and mathematically the most convenient one — is to require $F - F_\lambda$ to be small in the *uniform* operator topology, i.e., to require $\|P_T(F - F_\lambda)\|_\Delta$ to be small for all $T \in S$. This would then typically take into consideration errors in time constants, in gains, and in initial conditions. It is, however, *not* the choice adopted here since it is merely required that F should

---

[4] A basic paper on the well-posedness of feedback systems is that by Zames (Ref. 1). The setting considered in the present monograph is quite different, however. Reference 1 points out somewhat more carefully why the concept of "physical system" that is adopted here is a natural one. This is done in terms of ideal elements.

approximate F_λ in the *strong* operator topology, i.e., that $(F - F_\lambda)x$ should be small for any *fixed* x. The reason for adopting this choice, which makes the problem of well-posedness mathematically much more difficult, arises from engineering considerations. Indeed, examples of typical effects that are invariably ignored in the mathematical model are: small time delays, band limitation over broad bandwidths, distributed effects, small time constants, and similar phenomena that are believed to happen much faster than any changes due to the dominant dynamics of the system or in the expected variations in the inputs. It can easily be verified that none of these effects is (in general) continuous in the uniform operator topology. The induced norm on L_p^B, $1 \leqslant p \leqslant \infty$, of the identity minus a delay of length $\lambda > 0$, is always two. The same holds for the identity minus the system with transfer function $I/(\lambda s + 1)$, $\lambda > 0$, no matter how small λ, and thus no matter now closely the impulse response $(1/\lambda)e^{-t/\lambda}$ "resembles" a unit impulse. These effects are close to the identity, however, in the strong topology on L_p^B, $1 \leqslant p \leqslant \infty$, in the sense that $\lim_{\lambda \to 0} F_\lambda x = x$ for any $x \in L_p^B$, $1 \leqslant p \leqslant \infty$.

A careful examination of the definition of well-posedness also reveals that a *pure delay* T_ϵ, is assumed to be present in the loop describing the physical feedback system. This assumption is inspired by the consideration that no information can travel faster than the speed of light, which is a finite, albeit very large, constant. It thus seems very reasonable to make such an assumption. It should be noted that this assumption has not been introduced capriciously, but that it is essential to the mathematical development.

The conditions assumed on the physical model in the treatment of well-posedness appear to be minimal. Taking into consideration that the final conditions will be in terms of the mathematical models, *no* additional assumptions on the physical systems will be made. This is in the spirit that the exact description of the physical system is unknown but that smoothness assumptions can be made, these being warranted by general physical principles.

The assumption made is thus that the summing junctions and the systems in the forward and the feedback loop of the system are approximations (in the strong topology) of the exact physical (strongly causal and locally continuous) system. The requirement for well-posedness of the closed loop system is then that it itself approximate the physical system obtained by considering exact physical models for the summing junctions and the systems in the forward and feedback loop. Thus well-posedness of the open-loop operators (i.e., the open-loop operators are approximations of physical systems) should translate into well-posedness of the closed-loop system.

A final word of caution in the interpretation of the well-posedness condition is necessary. The requirement is *not* that

$$\lim_{\epsilon\downarrow 0,\lambda\to 0} (F_{1\lambda}^{\epsilon}(u_1,u_2), F_{2\lambda}^{\epsilon}(u_1,u_2)) = (I + G)^{-1}(u_1,u_2),$$

but precisely that

$$\lim_{\epsilon\downarrow 0}\lim_{\lambda\to 0} (F_{1\lambda}^{\epsilon}(u_1,u_2),F_{2\lambda}^{\epsilon}(u_1,u_2)) = (I + G)^{-1}(u_1,u_2).$$

It will be remarked later on that this unfortunate situation appears to be the best that can be obtained. This phenomenon is not unlike similar difficulties encountered in stochastic systems (Ito calculus) and differential games.

4.3.2 Examples of Ill-Posed Feedback Systems

A surprising fact concerning well-posedness of feedback systems is that the operators appearing in the loop need *not* be pathological to result in an ill-posed feedback system. A number of interesting examples of ill-posed feedback systems can be found in the literature (see Ref. 1, and further references therein). Three other examples are the following:

1. Let $G_1 = -I$ and $G_2 = I$. Then the feedback system has no solutions if $u_1 \neq u_2$ and has multiple solutions if $u_1 = u_2$.

2. Assume that the forward loop consists of a delay of length $T > 0$ minus a unit gain, that $G_2 = I$, and that $u_2 = 0$ (see Figure 4.4). The

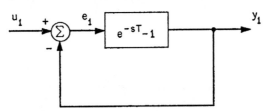

Figure 4.4   The Feedback System of Example 2

error is then (uniquely) given in terms of the input by $e_1(t) = u_1(t + T)$ for $t \geqslant T_0$ and thus depends on the input in an anticipatory way.

3. Assume that all signals in the loop are real-valued functions on $[T_0,\infty)$, that the forward loop consists of a constant gain K, that $G_2 = I$, and that $u_2 = 0$ (see Figure 4.5). If $K \neq -1$, then this feedback system has a unique solution $e_1(t) = (1 + K)^{-1}u_1(t)$ for any u_1. If a

small delay of length $\lambda > 0$ is introduced in the forward loop, then the error corresponding to a nonzero constant input $u_1(t) = u_0 = $ constant $\neq 0$ is given by

$$
e_1(t) = \begin{cases}
u_0 & \text{for } 0 \leqslant t < \lambda \\
u_0 - Ku_0 & \text{for } \lambda \leqslant t < 2\lambda \\
\vdots & \vdots \\
u_0 - Ku_0 + K^2u_0 - \cdots + (-K)^n u_0 & \text{for } n\lambda \leqslant t < (n+1)\lambda \\
\vdots & \vdots \\
\vdots
\end{cases}
$$

Hence if $\lambda \to 0$, $e(t)$ approaches $(1 + K)^{-1}u_0$ (the solution for $\lambda = 0$) *if and only if* $|K| < 1$ (when the expansion $(1 + K)^{-1} = \sum_{n=0}^{\infty} (-K)^n$ is valid). The result obtained in this example is characteristic of the conditions for well-posedness which will be obtained next.

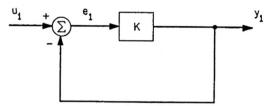

Figure 4.5   The Feedback System of Example 3

4.3.3 A Condition for Well-Posedness

Well-posedness imposes a weak condition on the operators in the feedback system. One such condition is given in the theorem that follows. First, however, a few more definitions.

Definitions: Let F be a locally Lipschitz-continuous causal operator from W_e into itself. Let $T \in S$ and $\Delta T > 0$ be given. Then the real number

$$
\rho_T(\Delta T) \triangleq \sup_{\substack{x, y \in W_e \\ P_T(x-y)=0 \\ P_{T+\Delta T}(x-y) \neq 0}} \frac{\|P_{T+\Delta T}(Fx - Fy)\|}{\|P_{T+\Delta T}(x - y)\|}
$$

is called the *gain of F on the interval* $[T, T + \Delta T]$. It is clear that $\rho_T(\Delta T)$ is monotone nondecreasing in ΔT for T fixed. The real number $\rho_T \triangleq \inf_{\Delta T > 0} \rho_T(\Delta T) = \lim_{\Delta T \downarrow 0} \rho_T(\Delta T)$ is sometimes referred to as the *instantaneous gain* of F at T. Note that $\rho_T = 0$ for a strongly causal operator.

Let F be a locally causal Lipschitz continuous operator from W_e into itself. Let $T \in S$ and $\Delta T > 0$ be given. Then there exist real numbers $K_T < \infty$ and $M_T < \infty$ such that

$$\|(P_{T+\Delta T} - P_T)(Fx - Fy)\| \leqslant K_T \|(P_T(x - y)\|$$
$$+ M_T \|(P_{T+\Delta T} - P_T)(x - y)\|$$

for all x, $y \in W_e$. The greatest lower bound of all real numbers M_T satisfying this condition for all x, $y \in W_e$ and some $K_T < \infty$ will be called[5] the *uniform gain of F on the interval* $[T, T + \Delta T]$. If F is linear then its gain and its uniform gain on the interval $[T, T + \Delta T]$ are equal. The definition of *uniform instantaneous gain* is now obvious. The following theorem states sufficient conditions for well-posedness.

THEOREM 4.1

The feedback system described by equations (FE) is well posed if either of the following conditions is satisfied for all $T \in S$:

1. The product of the uniform instantaneous gains of the operators G_1 and G_2 is less than $\alpha < 1$.
2. The system is linear and the instantaneous gain of the Nth power of the open-loop operator, $(G_2 G_1)^N$, is less than unity for some integer $N > 0$.

Proof: The proof is an adaptation of the usual proofs involving existence and uniqueness of solutions of ordinary differential equations or Volterra integral equations.[6] Since the interested reader will be

---

[5] Whether or not this terminology is natural is unclear. The idea of instantaneous gain and the gain on the interval $[T, T + \Delta T]$ is obvious and clearly related to the feedthrough in a system. The uniform instantaneous gain and the uniform gain on the interval $[T, T + \Delta T]$ are also related to the feedthrough in a system but in addition take into consideration that the state of the system could be different.

[6] The type of proof given here can be completely imbedded in a contraction-mapping argument, thus avoiding the argumentation on consecutive intervals. It involves redefining the norm of $P_{t_0+n\Delta T}x$ to the equivalent norm

$$\|P_{t_0+n\Delta T}x\| \triangleq \sum_{i=1}^{n} \rho_i^2 \|(P_{t_0+i\Delta T} - P_{t_0+(i-1)\Delta T})x\|,$$

with ρ_i appropriately chosen nonzero constants. Both arguments are tedious and painful. The one given here appears to be the more transparent one. The contraction-mapping argument is somewhat more sophisticated and yields an important side result: it shows that the successive approximations $e_{n+1} = -Ge_n + u$ converge on any interval $[t_0, t_0 + T]$ to the solution and it gives a (more or less) explicit bound on the rate of convergence and on the error. Similar ideas can be found in References 2, 3.

quite familiar with these arguments, let a somewhat sketchy proof therefore suffice.

Consider first case 1. The feedback equations (FE) can be written as $e = -Ge + u$ with $u = (u_1, u_2)$, $e = (e_1, e_2)$, and $Ge \triangleq (G_2 e_2, -G_1 e_1)$. Consider the product space $W_1 \times W_2$ and its extension $W_{1e} \times W_{2e}$. Let $\alpha_1 \triangleq \|P_{T_0 + \Delta T} G_2 P_{T_0 + \Delta T}\|_\Delta$ and $\alpha_2 \triangleq \|P_{T_0 + \Delta T} G_1 P_{T_0 + \Delta T}\|_\Delta$. Let $\Delta T > 0$ be such that $\alpha_1 \alpha_2 < 1$. Such a $\Delta T > 0$ exists by the assumptions of the theorem. Consider now the equivalent norm on $W_1 \times W_2$ defined by

$$\|x\| = \|(x_1, x_2)\| \triangleq \beta_1 \|x_1\|_{W_1} + \beta_2 \|x_2\|_{W_2},$$

with $\alpha_1^{-1} > \beta_2 \beta_1^{-1} > \alpha_2$. Such a β_1 and β_2 exist since $\alpha_1 \alpha_2 < 1$. A simple calculation now shows that $P_{T_0 + \Delta T} G P_{T_0 + \Delta T}$ is a contraction on $W_{1(T_0, T_0 + \Delta T_0)} \times W_{2(T_0, T_0 + \Delta T)}$. It thus follows that the operator $P_{T_0 + \Delta T}(I + G) P_{T_0 + \Delta T}$ is invertible on the spaces restricted to the interval $[T_0, T_0 + \Delta T]$. The argument is then repeated on consecutive intervals, and it follows from continuity that these intervals will cover the whole half-line $[T_0, \infty)$. This argument thus yields existence and uniqueness of solutions and condition W.1 for well-posedness. Causality of the inverse follows from the fact that the invertibility is based on the convergence of the successive approximations $e_{n+1} = Ge_n + u$, $n \in I^+$, with e_0 arbitrary. This approximation technique then yields causality of the inverse when used together with the above construction of the solution. Condition WP.2 for well-posedness hence follows. Notice the conditions WP.1 and WP.2 did *not* use the uniformity of the gain. The argument given here thus also proves Theorem 2.16. Condition WP.3 follows readily from this uniformity and the successive approximations on consecutive intervals. To show condition WP.4, observe that the first part of this theorem yields, for $\epsilon < 0$ sufficiently small, the existence, causality, and Lipschitz continuity of the operators $F_{1\lambda}^\epsilon$ and $F_{2\lambda}^\epsilon$ introduced in condition WP.4.

To show that $(I + G)^{-1}(u_1, u_2) = \lim_{\epsilon \uparrow 0} \lim_{\lambda \to 0} (F_{1\lambda}^\epsilon(u_1, u_2), F_{2\lambda}^\epsilon(u_1, u_2))$ one proceeds as follows: Let $-\epsilon > 0$ be fixed and consider the operator G_ϵ defined by $G_\epsilon(e_1, e_2) = (-T_\epsilon G_2 e_2, G_1 e_1)$. Let $e_1 \in W_{1e} \times W_{2e}$ and consider now the successive approximations obtained by defining $e_{n+1} = -G_\epsilon e_n + u$ and $e_{n+1}^\lambda = F(e_n)$, where F is the operator on the right-hand side of the equation (PFE). By the strong convergence (assumption 4 of condition WP.4) it follows that for all $n \in I^+$, $\lim_{\lambda \to 0} e_{\lambda n}^\epsilon = e_n^\epsilon$. Moreover, $\lim_{n \to \infty} e_{n\lambda}^\epsilon = e_\lambda^\epsilon$ and $\lim_{n \to \infty} e_n^\epsilon = e^\epsilon$, and these limits are uniform since $-\epsilon > 0$ (in fact, all series involved are really finite series). Thus $\lim_{\lambda \to 0} e_\lambda^\epsilon = e^\epsilon$. Comparing now the solutions of

$e^\epsilon = -G_\epsilon e^\epsilon + u$ and $e = -Ge + u$, one observes that they also can be obtained by successive approximations and that the convergence is, by the fact that the product of the instantaneous gains is less than unity, uniform in ϵ. Thus $\lim_{\epsilon \downarrow 0} e^\epsilon = e$, as claimed. This procedure is then continued on the successive intervals. It remains to be shown that the operators $G_{1\lambda}F^\epsilon_{1\lambda}$ and $T_\epsilon G_{2\lambda}F^\epsilon_{2\lambda}$ are strongly causal. This, however, is immediate since for all $-\epsilon > 0$ there is a delay in the loop. This ends the proof for the nonlinear case.

The refinement for linear systems follows essentially the same lines as the nonlinear case but with the contraction mapping principle replaced by Cacciopoli's corollary (Ref. 4). The same recursive algorithm works, and the argument only changes in detail. These details are left to the reader.

4.3.4 Discussion of the Well-Posedness Condition

Theorem 4.1 imposes a restriction on the amount of feedthrough in G. Some important particular cases are given in Corollaries 4.1.1–4.1.3.

Definition: Let F be a causal operator from $W_{1\epsilon}$ into itself. Then F is said to *delay* all inputs if for some $\epsilon > 0$ the operator F_ϵ defined as $F_\epsilon x(t) \triangleq Fx(t + \epsilon)$ is also causal (i.e., F can be cascaded with a predictor and the composition remains causal).

COROLLARY 4.1.1

The feedback system described by the equations (FE) is well posed if the open loop operator $G_2 G_1$ delays all inputs.

Proof: This corollary follows from Theorem 4.1.

COROLLARY 4.1.2

The feedback system described by the equations (FE) is well posed if the operator G_1 is strongly causal, uniformly with respect to past inputs.

Proof: This corollary follows from Theorem 4.1 and the fact that G_2 is locally Lipschitz continuous.

An important example of a strongly causal operator is when the output y is related to the input u through the ordinary differential equation

$$\dot{x}(t) = f(x(t), u(t), t)$$
$$y(t) = g(x(t), t)$$

with $t \geqslant t_0$ and $x(t = t_0)$ given, where $x \in B_x$, $u \in B_u$, and $y \in B_y$ with B_x, B_u, and B_y given Banach spaces, and f and g are continuous in t for $t \geqslant t_0$ and Lipschitz continuous on respectively $B_x \times B_u$ and B_x. It is easy to show that G defined by $y \triangleq Gu$ is then a well-defined strongly causal operator, uniformly with respect to past inputs, mapping from $L_{pe}^{B_u}$ into $L_{pe}^{B_v}$, $1 \leqslant p \leqslant \infty$.

Definition: Let g be an operator from $B \times [T_0, \infty)$ into itself. Then the operator G defined by $y(t) \triangleq g(u(t), t)$ will be called the *memoryless operator with characteristic g*.

COROLLARY 4.1.3

Let $W_1 = L_p^{B_1}(T_0, \infty)$, $1 \leqslant p \leqslant \infty$, and $G_2 G_1 = G' + G''$ be linear, with G' a memoryless operator and G'' strongly causal, uniformly with respect to past inputs. Assume that the characteristic of G', g, is continuous in t on $[T_0, \infty)$ and continuous on B. Then the feedback system described by the equations (FE) is well posed if g satisfies for some $\alpha < 1$ and all t in any finite interval the inequality

$$\lim_{N \to \infty} \sup_{\substack{u_1 \in B_1 \\ u_1 \neq 0}} \left(\frac{\|g^N(u_1, t)\|^{1/N}}{\|u_1\|} \right) \leqslant \alpha < 1.$$

Conversely, if for some integer $N > 1$, some element $u_1 \in B_1$, some $t \geqslant t_0$, and some λ, $|\lambda| \geqslant 1$, the characteristic satisfies $g^N(u_1, t) = \lambda u_1$, then the feedback system described by the equations (FE) is ill posed.

Proof: The well-posedness part is a simple consequence of Theorem 4.1. The ill-posedness part evolves along the lines of Example 3 of Section 4.3.2. Details of the proof can be found elsewhere (see Ref. 5).

Well-posedness thus results if the spectral radius of g is less than unity, and ill-posedness results if g_1 has an eigenvalue outside on the unit circle for some $t \geqslant t_0$. In particular, if B_1 is a finite-dimensional space, then the condition that all the eigenvalues of g be inside the unit circle for all $t \geqslant t_0$ becomes a *necessary and sufficient* condition for well-posedness. It can be shown, moreover, that the decomposition of $G_2 G_1$ as given in Corollary 4.1.3 is canonical for *linear* systems.

Remark 1: The question of well-posedness on the extended space when the time-interval of definition is $(-\infty, +\infty)$ is much more intricate and is intimately related to the continuity of the feedback system. This aspect will be further discussed in Section 4.6.[7]

[7] See also Reference 6. The author is *not* convinced, however, that the study of feedback systems with the time-interval of definition $(-\infty, +\infty)$ rests on physical grounds.

The result obtained in Theorem 4.1 has important implications for the engineer who is modeling a feedback system. Since it is *not* always desirable to model with extreme accuracy, it often happens that one likes to model the open loop with a feedthrough. Theorem 4.1 then gives very specific constraints on this feedthrough for well-posedness. These conditions are, moreover, all but necessary and sufficient for linear systems. If the conditions are not satisfied and the feedthrough is too large, then the feedback system will most likely be ill-posed and the mathematical model should be modified. This will probably most easily be done by taking into consideration some higher-order effects until the desired limitation on the feedthrough is obtained. This will of course generally augment the complexity of the system, in particular the dimension of the state space.

Remark 2: One of the conclusions of well-posedness states that

$$\lim_{\epsilon \downarrow 0} \lim_{\lambda \to 0} (F_{1\lambda}^{\epsilon}(u_1,u_2), F_{2\lambda}^{\epsilon}(u_1,u_2)) = (I + G)^{-1}(u_1,u_2),$$

and the limits on the left-hand side are, somewhat unfortunately, to be taken *in that specific order*. An example which shows the necessity of this unfortunate situation is a feedback system with the gain $k(t) = 1$ for $T_0 \leqslant t \leqslant T_0 + \lambda$, $\lambda > 0$, and zero otherwise, a delay of length ϵ in the forward loop, and a unit feedback. It is then clear that if $\lambda \to 0$ and $\epsilon \to 0$, the feedback system — albeit well posed — yields a non-existent limit for $\lim_{\epsilon \downarrow 0} (F_{1\lambda}^{\epsilon}(u_1,u_2), F_{2\lambda}^{\epsilon}(u_1,u_2))$ for $\lambda > 0$. Notice also that this difficulty remains when the inputs are assumed to be smooth.

4.4 Stability and Instability

From a mathematical point of view, stability is analysis: given a mathematical equation, one bounds certain quantities. Optimal control, on the other hand, is synthetic in nature: an unknown is to be determined in some optimal fashion. Notwithstanding these analytical aspects, stability theory has been very successfully used as the basis for the synthesis of feedback controllers.[8] In fact, *essentially all the classical design techniques are directly inspired by stability considerations.*

Modern control theory tends to put heavy emphasis on optimal control and with it on general optimization theory — the derivation of

---

[8] Similar synthesis techniques should be developed using modern stability theory concepts. This has not been done as yet although it could lead to successful design techniques for nonlinear and/or time-varying systems.

necessary conditions for optimality, and solution techniques based on these, usually in the form of algorithms. Does this mean that stability theory is passé or, at best, a side issue in modern control theory? It appears not. Mathematical optimization theory allows only in the rarest circumstances for an exact mathematical design of a feedback controller, and, even if this were possible, such a controller is then almost never implemented, be it because too many sensors are required or because of some other technical difficulty. Thus, designs based on optimization theory usually resort to a simplified mathematical model and deal with sometimes crude approximations at all stages of the development. It should also be recognized that stability is often the main concern in control, rather than optimality with respect to a (sometimes somewhat arbitrary) mathematical performance criterion. In other words, the main issue in many control problems is to guarantee that a plant will operate in the neighborhood of a desired operating point in the face of disturbances. Precisely how the plant returns to this operating point after a disturbance is quite often of secondary importance. Such control problems pose two challenges to the designer: 1, an *optimization problem* (often a nondynamical optimization to determine the most efficient steady-state operation, or an open-loop dynamic optimal control problem to determine the optimal transfer); and 2, a *stability problem* (to guarantee that the closed loop control will keep the system variables around their optimal values).

Two distinct approaches to the problem of stability of systems can be taken. The first — and more traditional one — regards stability as an *internal* property of a system: the system is considered as excited by an initial condition, and boundedness or convergence of the state for future time is taken as the basic requirement for stability. The second approach — and, from a modern system theory point of view, a more logical one — regards stability as an *input-output* property: the system is regarded as a mapping between normed spaces, and boundedness of this map is taken as the basic requirement for stability.

Input-output stability is, from an engineering point of view, a very significant and important type of stability. The informal definition of stability given for instance by Nyquist in his classic paper on "Regeneration Theory" (Ref. 7) is essentially that of input-output stability. It is intimately related to the idea of "stability under constant disturbances" and thus has some classical — although not system-oriented — foundations. The concept of input-output stability stands in direct competition with the idea of stability in the sense of Lyapunov.

Input-output stability considers as the disturbance entering the system a constantly acting input, whereas stability in the sense of Lyapunov considers the initial conditions as the disturbance in the system. Which of these two types of stability is to be preferred clearly depends on the particular application. In a sense, input-output stability protects against noise disturbances whereas Lyapunov stability protects against a single impulselike disturbance.

As is apparent from the definition formally introduced in Section 4.2, the particular state space is not relevant in the concept of input-output stability and usually does *not* enter into the development of stability conditions in an essential way. The simplicity of the results usually depends — besides on the system itself — on the input and output spaces. These could of course be very simple, e.g., consisting of scalar functions of time, even though the state space and the state transition process could be very complex. Many systems encountered in engineering applications do indeed have this property: there are few output sensors and few control manipulators, but the state space and the state transition process are very complex and governed, for example, by a partial differential equation. This particular aspect of input-output stability makes this concept particularly appealing when studying the stability of distributed-parameter systems and, more specifically, those described by partial differential equations where difficulties of a theoretical nature generally limit the applicability of Lyapunov-based methods.

Lyapunov stability[9] considers stability as an internal property of a system, and inputs and outputs do not play a role. This formulation accounts for the early development and great historical importance of this type of stability. The study of systems without inputs and outputs is indeed basic to classical dynamics. The traditional question of the stability of the solar system, for example, remains a long-standing challenge and does not involve inputs in any way. It is thus more than natural that stability of control systems has been studied in this context; namely, as a condition on undriven classical "dynamical systems."

[9] Most of the results presented here have their analogue obtained using Lyapunov methods. The survey paper by Brockett (Ref. 8) contains one such method based on spectral factorization and ample references to other works in this area. The same author presents in Reference 9 another method using algebraic matrix equations to obtain similar results. This method was originally developed by Popov (Ref. 10) and Kalman (Ref. 11). A survey paper by the author (Ref. 12) contains further references to the literature in this area, particularly those applying to distributed parameter systems as well.

This is in spite of the fact that its founders, Lyapunov and Poincaré, were not primarily interested in control. It should be noted that this dynamical-system point of view is supported by the work of Maxwell and much of the subsequent work on regulators. Although Lyapunov stability remains important and very useful in many control applications, its basic philosophy can often be challenged and is somewhat out of line with the modern approach to systems, where inputs and outputs are the fundamental variables and the state is merely an auxiliary variable that essentially represents the contents of a memory bank. The development and success of input-output stability should thus come as no surprise. This does not exclude that for many applications (for example, in aerospace problems) Lyapunov stability does represent a very satisfactory type of stability, and thus its study will remain both important and fruitful.

4.4.1 *Discussion of the Definition of Stability*

Although the concept of *input-output stability* (the type of stability studied here) is relatively old, its actual development is of a rather recent date. The basic idea of this type of stability is simple: the system is considered as a mapping between normed spaces — the input space and the output space — and boundedness of this map is taken as the basic requirement for stability. This boundedness of the input-output mapping then yields a bound on the norm of the output in terms of the norm of the input.

This informal definition of input-output stability uncovers on close examination a basic difficulty: it is not a priori clear what to take for the output space. More specifically, suppose that the output, y, is given in terms of the input, u, through the mapping $y = Fu$. If u belongs to a normed space U, then input-output stability roughly requires that y belong to a normed space Y. The difficulty is that it is not a priori clear whether the output $y = Fu$ will even be defined for all $u \in U$. This could of course be added as an additional requirement for stability; i.e., it could be understood that the definition of stability requires that $u \in U$ imply that $y = Fu \in Y$. Very often, however, the output $y = Fu$ is actually well defined for all $u \in U$ (as an element of some larger function space), even when the system is not input-output stable. This possibility of extending the basic input and output spaces is the underlying idea behind the introduction of extended spaces and is appropriate when F is causal. The introduction of extended spaces is the key fact which has led to the very successful application of functional methods in stability theory. Extended spaces were in fact first introduced in this context.[10]

The preceding remarks hold — amplified — when the input and the output are related through an implicit equation. Such implicit equations describing the input-output behavior occur specifically when the system is of a feedback type and are hence of more interest in stability analysis, since control naturally leads to a feedback configuration and stability problems. Recall that all the operators in the feedback system described by the equations (FE) are assumed to be causal.

Causality is a fundamental property of physically realizable systems, since it merely expresses that past and present output values do not depend on future input values. There is, however, another fundamental reason for treating input-output stability in the context of causal systems only. This reason is that stability requires some type of convergence as $t \to \infty$ and thus, unless the past and future play essentially different roles in the original system, such a definition, which reflects the future behavior only, would appear to be ill founded.

Considering now the actual definition of stability as given in Section 4.2, observe that it requires well-posedness and some type of boundedness of the outputs in terms of the inputs. The first condition is not really necessary and one could very well define and study stability *without* reference to well-posedness. Although such an approach does not violate any mathematical principles, two points should be kept in mind:

1. Well-posedness is a much more fundamental requirement than stability and should always be verified anyway.
2. When a feedback system is not well posed, it does not adequately describe the physical system it attempts to model. To correct this situation one will thus have to modify the mathematical model somewhat and *such a modification could very well alter some fundamental properties of the feedback system as, for instance, its stability*

---

[10] The idea of introducing the concept of extended spaces is due to Sandberg (Ref. 13) and Zames (Ref. 14). The subsequent development of input-output stability is largely due to the same authors. Survey articles summarizing their results are References 15, 16. The last reference is particularly valuable since it bases its rigorous analysis on a great deal of intuitive and physical reasoning. The approach followed here is presented in Ref. 6. The idea of extended spaces has appeared implicitly in the mathematical literature in the context of ordinary and partial differential equations, Volterra integral equation, and delay-differential equations. Hopefully, this concept will eventually be formally introduced in the mathematical literature for its elegance in treating causal operators and continuations as the one involved in the proof of Theorem 4.1.

properties.[11] It is thus natural to make well-posedness an a priori requirement for stability.

The second ingredient in the definition of stability is the requirement that inputs in *nonextended* spaces should lead to outputs and errors which are also in *nonextended* spaces. Thus "small" inputs should generate "small" outputs and errors. Notice also that this boundedness singles out the zero solution as the desired solution and ensures that a "small" noise driving the system will generate a correspondingly "small" output. It is thus the natural condition to impose on systems which are regulating the output around a *fixed* reference (taken without loss of generality to be zero).

4.4.2 Conditions for Stability

In this section some general and some specific conditions for stability are derived. The first theorem states a general result which is merely a rephrasing of definitions.

THEOREM 4.2 (STABILITY)

Consider the feedback system described by the equations (FE) and let G map $W_{1e} \times W_{2e}$ into itself according to $G(e_1, e_2) = (G_2 e_2, -G_1 e_1)$. Then the feedback system is stable if and only if:

1. It is well posed.
2. The inverse $(I + G)^{-1}$ (which exists on $W_{1e} \times W_{2e}$ by condition 1) maps $W_1 \times W_2$ into itself.

The system is finite gain stable if and only if in addition

3. The inverse $(I + G)^{-1}$ is bounded on $W_1 \times W_2$.

Proof: The definition of stability and Theorem 2.1 lead immediately to this result.

Theorem 4.2 leads to the following interesting alternate definitions of stability.

Alternate definition of stability (I). The feedback system described by equations (FE) is *finite gain stable* if and only if:

ST.1. It is well posed.
ST.2. For any inputs $u_i \in W_{ie}, i = 1, 2$, there exist constants $\rho_1, \rho_2 < \infty$ (independent of the inputs and T) such that for all $T \in S$, $\|P_T e_1\|$, $\|P_T e_2\|$, $\|P_T y_1\|$, $\|P_T y_2\| \leqslant \rho_1 \|P_T u_1\| + \rho_2 \|P_T u_2\|$.

[11] This follows, for instance, from the Nyquist Criterion and the Circle Criterion.

Alternate[12] *definition of stability* (II). The feedback system described by equations (FE) is *finite gain stable* if and only if:

ST.1. It is well posed.

ST.2. For any inputs $u_i \in W_{ie}$ with

$$\sup_{T \geqslant T_0} \frac{1}{T - T_0} \|P_T u_i\| < \infty, \, i = 1, 2,$$

there exist constants $\rho_1, \rho_2 < \infty$ (independent of their inputs and T) such that

$$\sup_{T \geqslant T_0} \frac{1}{T - T_0} \|P_T e_1\|, \, \sup_{T \geqslant T_0} \frac{1}{T - T_0} \|P_T e_2\|,$$

$$\sup_{T \geqslant T_0} \frac{1}{T - T_0} \|P_T y_1\|, \, \sup_{T \geqslant T_0} \frac{1}{T - T_0} \|P_T y_2\|$$

$$\leqslant \rho_1 \sup_{T \geqslant T_0} \frac{1}{T - T_0} \|P_T u_1\| + \rho_2 \sup_{T \geqslant T_0} \frac{1}{T - T_0} \|P_T u_2\|.$$

It can be argued successfully that the alternate definition (I) reflects a more fundamental boundedness of the response and should hence (although being equivalent to the original definition) be taken as the basic definition of stability. Definition (I) indeed expresses a bound on the response in terms of the inputs but allows for a larger and a *somewhat more realistic* class of testing inputs. This appears to be an important point since most noise inputs fall in the latter class and stability thus indeed expresses some boundedness of the response to such inputs.

Theorem 4.2 leads to the following theorem for stability in terms of the open loop operator.

THEOREM 4.3

Consider the feedback system described by the equations (FE) and assume that:

1. It is well posed.
2. The operator G_2 is Lipschitz continuous on W_2.
3. The operator G_1 is bounded on W_1 or there exists an $\epsilon > 0$ such that for all $x \in W_{2e}$ and $T \in S$, $\|P_T G_2 x\| \geqslant \epsilon \|P_T x\|$.

[12] This formulation has interesting applications to problems in which the inputs are described in terms of random processes.

Then $I + G_2G_1$ is invertible on W_{1e} and the feedback system is finite gain stable if and only if $\|(I + G_2G_1)^{-1}\| < \infty$ on W_1.

Proof: The operator $I + G_2G_1$ has a causal inverse on W_{1e} by well-posedness. Let $u_1 \in W_1$ and $u_2 \in W_2$ be given, and let $e_1 \in W_{1e}$ be the corresponding error. Then $e_1 + G_2G_1e_1 = u_1 + G_2G_1e_1 - G_2(G_1e_1 + u_2)$. From the Lipschitz condition on G_2 it thus follows that for all $T \in S$, $\|P_T(I + G_2G_1)e_1\| \leqslant \|P_Tu_1\| + \|G_2\|_\Delta \|P_Tu_2\|$. Since $(I + G_2G_1)^{-1}$ is causal and bounded on W_{1e},

$$\|P_Te_1\| \leqslant \|(I + G_2G_1)^{-1}\| \, \|P_T(I + G_2G_1)e_1\|.$$

Thus $\|P_Te_1\| \leqslant \|(I + G_2G_1)^{-1}\| \, \|u_1\| + \|(I + G_2G_1)^{-1}\| \, \|G_2\|_\Delta \, \|u_2\|$. This shows that $e_1 \in W_1$ and that the finite gain condition is indeed satisfied. Since $e_1 = u_1 - y_2$, this yields immediately that y_2 also satisfies these conditions. It remains to be shown that e_2 and $y_1 \in W_2$ and that their gain conditions are satisfied. If $\|G_1\| < \infty$, this follows immediately. If the second condition in assumption 3 is satisfied, then it follows that $\epsilon \|P_Te_2\| \leqslant \|P_Ty_2\|$, which shows that $e_2 \in W_2$ and that $\|e_2\| \leqslant \epsilon^{-1} \|y_2\|$. From $e_2 \in W_2$ and $e_2 = u_2 + y_1$, it then follows that y_1 also satisfies such conditions. The necessity part of the theorem is obvious by letting $u_2 = 0$ and applying Theorem 4.2.

Notice that under the assumptions of Theorem 4.3 finite gain stability results *if and only if* there exists a real number $\epsilon > 0$ such that for all $T \in S$ and $x \in W_1$, $\|P_T(I + G_2G_1)x\| \geqslant \epsilon \|P_Tx\|$.

It is now a simple matter to use the invertibility theorems of Chapter 2 to obtain more specific conditions for stability.[13] These conditions are stated in the following corollaries.

COROLLARY 4.3.1

The feedback system described by the equations (FE) is finite gain stable it is well posed, if G_2 is Lipschitz continuous on W_2, if G_1 is bounded on W_1, and if the open loop is attenuating on W_1 (i.e., if $\|G_2G_1\| < 1$).

COROLLARY 4.3.2

Let $W_1 = W_2$. Then the feedback system described by the equations (FE) is finite gain stable if it is well posed, if G_2 is Lipschitz continuous

---

[13] These results are essentially due to Zames (Refs. 14, 16) and Sandberg (Refs. 13, 15). The only novel part in the Corollaries as presented here is that the operators G_2 and G_1 are never separated (compare the $\|G_2G_1\| < 1$ conditions with the $\|G_2\| \, \|G_1\| < 1$ condition of Ref. 16). A theorem of the positive-operator type has appeared in Ref. 17 in a Lyapunov setting.

on W_2, and if there exists a scalar c such that $I + cG_1$ has a causal inverse on W_{1e} such that $(I + cG_1)^{-1}$ is bounded on W_1 and $(G_2 - cI)G_1(I + cG_1)^{-1}$ is attenuating on W_1.

COROLLARY 4.3.3

Let $W_1 = W_2$ be a Hilbert space. Then the feedback system described by equations (FE) is finite gain stable if it is well posed, if G_2 is Lipschitz continuous on W_2, and if for some real numbers $a \leqslant b$, $b > 0$, the operator G_2 is strictly inside the sector $[a,b]$, $I + \frac{1}{2}(a + b)G_1$ has a causal inverse on W_{1e}, and G_1 satisfies one of the following conditions:

1. $a < 0$, and G_1 is inside the sector $[-1/b, -1/a]$ on W_1;
2. $a > 0$, and G_1 is outside the sector $[-1/a, -1/b]$ on W_{1e}; or
3. $a = 0$, and $G_1 + (1/b)I$ is positive on W_{1e}.

COROLLARY 4.3.4

Let $W_1 = W_2$ be a Hilbert space. Then the feedback system described by equations (FE) is finite gain stable if it is well posed, if G_2 is positive on W_{2e}, and if G_1 is strictly positive and Lipschitz continuous on W_1.

Proof: These corollaries follow from Theorem 4.3 and Theorem 2.17, Theorem 2.18, Corollary 2.20.1, and Theorem 2.21 (occasionally with the roles of G_2 and G_1 reversed). The details are left to the reader.

Remark 1: It goes without saying that the roles of G_1 and G_2 can be reversed in the preceeding theorems. These theorems also remain valid if the conicity or positivity conditions are stated in terms of a causal factorization of G_2G_1 (i.e., in terms of causal operators G_2' and G_1' such that $G_2'G_1' = G_2G_1$ on W_1).

Remark 2: Notice that Corollaries 4.3.1 and 4.3.4 are very intuitive in nature. Corollary 4.3.1 states the stability results if the open loop attenuates all signals, and Corollary 4.3.4 states that stability results if the feedback system can be modeled as the parallel interconnection of passive systems (see Section 2.8.3).

4.4.3 Conditions for Instability

In this section a specific condition for instability is given. It is in a sense the converse of Corollary 4.3.2.

THEOREM 4.4 (INSTABILITY)

Consider the feedback system described by equations (FE) and assume that it is well posed and that G_2G_1 is Lipschitz continuous on

W_1. Let W_1' and G' be a backwards extension of W_1 and G_2G_1 from $S = [T_0, \infty)$ to $S' = (-\infty, +\infty)$, and assume that G' is Lipschitz continuous on W_1' and that for all $T \in S'$, the operator $I + G'$ has a causal inverse on $W_{1T_e}' \triangleq \{x \in W_{1e}' \mid P_T x = 0\}$. Let G_1', G_2' be Lipschitz continuous causal operators on W_1' and $I + G' = I + G_2'G_1'$. Then the feedback system is unstable if there exists a scalar c such that $I + cG_1'$ has a noncausal Lipschitz-continuous inverse on W_1' and if $(G_2' - cI)G_1'(I + cG_1')^{-1}$ is a contraction on W_1'.

Proof: Theorem 2.23 shows that $(I + G_2G_1)^{-1}$ (the inverse on W_{1e}) does *not* map W_1 into itself. Thus there is at least one input $u_1 \in W_1$ and $u_2 = 0$ such that $e_1 \in W_{1e} - W_e$ which thus yields instability.

COROLLARY 4.4.1

Let W_1 be a Hilbert space, and let the preliminary conditions of Theorem 4.4 be satisfied. Then the feedback system is unstable if for some real numbers $0 < a \leqslant b$, the operator G_2' is incrementally strictly inside the sector $[a,b]$ on W_1', $I + \frac{1}{2}(a + b)G_1'$ has a noncausal Lipschitz continuous inverse on W_1', and G_1' is outside the sector $[-1/b, -1/a]$ on W_1'.

Proof: This corollary follows the ideas explained in Section 2.8 in making the contraction condition of Theorem 4.4 more explicit.

The above results can be improved somewhat through linearization. This will be discussed in detail in Chapter 7.[14]

4.5 Continuity and Discontinuity

This section is concerned with the continuity and discontinuity of the feedback system. The definitions of continuity, Lipschitz continuity, and discontinuity have been given in Section 4.2. Much of the preliminary discussions of stability in Section 4.4 carries over to continuity.

The study of continuity of feedback systems is not standard in the engineering literature.[15] It is, however, an important concept for

[14] Related instability criteria (in a much more restricted context) have first been demonstrated in a Lyapunov setting by Brockett and Lee (Ref. 18). In their present setting and generality they originate with the author (Ref. 6).

[15] Continuity of feedback systems has first been explored by Zames (Ref. 16). The concept as such does not appear to have had a great deal of success, although it makes good sense (more so than stability) as a general nonexplosiveness condition for tracking systems.

systems for which the desired output is not a priori fixed (in other words, for tracking systems). The feedback system should then be viewed as follows: An (a priori undetermined) signal drives a stable feedback system and generates a desired output. Additive in this input is an undesired "small" noise component. Continuity then requires the corresponding output to differ from the desired output also by a "small" amount.

4.5.1 Conditions for Continuity

The first theorem on continuity is merely a rephrasing of definitions.

THEOREM 4.5 (CONTINUITY)

Consider the feedback system described by the equations (FE) and let G map $W_{1e} \times W_{2e}$ into itself according to $G(e_1, e_2) \triangleq (G_2 e_2, -G_1 e_1)$. Then the feedback system is continuous if and only if:

1. It is well posed.
2. The inverse $(I + G)^{-1}$ (which exists on $W_{1e} \times W_{2e}$ by 1) maps $W_1 \times W_2$ into itself and is continuous on $W_1 \times W_2$.

The system is Lipschitz continuous if and only if in addition:

3. The inverse $(I + G)^{-1}$ is Lipschitz continuous on $W_1 \times W_2$.

Proof: The definition of continuity and Theorem 2.1 lead immediately to this result.

Theorem 4.5 leads to the following interesting alternate definitions of continuity.

Alternate Definition of Continuity (I). The feedback system described by equations (FE) is (Lipschitz) continuous if and only if:

C.1. It is well posed.
C.2. Let $u_i \in W_{ie}$, $i = 1, 2$, be given (but arbitrary), and let e_i and y_i be the corresponding errors and outputs. Then there exists, given any $\epsilon > 0$, a $\delta > 0$ such that the inputs $u_i + \Delta u_i$ with $\Delta u_i \in W_i$ and $\|\Delta u_1\|$, $\|\Delta u_2\| \leq \epsilon$ yield corresponding errors, $e_i + \Delta e_i$, and outputs, $y_i + \Delta y_i$, with $\Delta e_1, \Delta y_2 \in W_1$, $\Delta e_2, \Delta y_1 \in W_2$, and $\|\Delta e_1\|$, $\|\Delta e_2\|$, $\|\Delta y_1\|$, $\|\Delta y_2\| \leq \delta$ (for some fixed $K < \infty$, the inequalities hold with $\delta = K\epsilon$).

Alternate Definition of Continuity (II). The feedback system described by equations (FE) is *Lipschitz continuous* if and only if:

C.1. It is well posed.

C.2. There exist constants ρ_1, $\rho_2 < \infty$ such that any $u_i^{(1)}$, $u^{(2)} \in W_{ie}$, $i = 1, 2$, yield corresponding errors, $e_i^{(1)}$, $e_i^{(2)}$, and outputs, $y_i^{(1)}$, $y_i^{(2)}$ satisfying $\|P_T(e_1^{(1)} - e_1^{(2)})\|$, $\|P_T(e_2^{(1)} - e_1^{(2)})\|$, $\|P_T(y_1^{(1)} - y_1^{(2)})\|$, $\|P_T(y_2^{(1)} - y_2^{(2)})\| \leqslant \rho_1 \|P_T(u_1^{(1)} - u_1^{(2)})\| + \rho_2 \|P_T(u_2^{(1)} - u_2^{(2)})\|$ for all $T \in S$.

Alternate Definition of Continuity (III). The feedback system described by equations (FE) is *Lipschitz continuous* if and only if:

C.1. It is well posed.

C.2. There exist constants ρ_1, $\rho_2 < \infty$ such that for any $u_i^{(1)}$, $u_i^{(2)} \in W_{ie}$ with

$$\sup_{T \geqslant T_0} \frac{1}{T - T_0} \|P_T(u_i^{(1)} - u_i^{(2)})\| < \infty, \qquad i = 1, 2,$$

the corresponding errors and outputs $e_i^{(1)}$, $e_i^{(2)}$ and $y_i^{(1)}$, $y_i^{(2)}$ satisfy

$$\sup_{T \geqslant T_0} \frac{1}{T - T_0} \|P_T(e_1^{(1)} - e_1^{(2)})\|, \sup_{T \geqslant T_0} \frac{1}{T - T_0} \|P_T(e_2^{(1)} - e_2^{(2)})\|,$$

$$\sup_{T \geqslant T_0} \frac{1}{T - T_0} \|P_T(y_1^{(1)} - y_1^{(2)})\|, \sup_{T \geqslant T_0} \frac{1}{T - T_0} \|P_T(y_2^{(1)} - y_2^{(2)})\|$$

$$\leqslant \rho_1 \sup_{T \geqslant T_0} \frac{1}{T - T_0} \|P_T(u_1^{(1)} - u_1^{(2)})\|$$

$$+ \rho_2 \sup_{T \geqslant T_0} \frac{1}{T - T_0} \|P_T(u_2^{(1)} - u_2^{(2)})\|.$$

THEOREM 4.6

Consider the feedback system described by the equations (FE) and assume that:

1. It is well posed
2. The operator G_2 is Lipschitz continuous on W_2
3. The operator G_1 is Lipschitz continuous on W_1 or there exists an $\epsilon > 0$ such that for all $x_1, x_2 \in W_{2e}$ and $T \in S$, $\|P_T(G_2 x_1 - G_2 x_2)\| \geqslant \epsilon \|P_T(x_1 - x_2)\|$.

Then $I + G_2G_1$ is invertible on W_{1e} and the feedback system is Lipschitz continuous if and only if $\|(I + G_2G_1)^{-1}\|_\Delta < \infty$ on W_1.

The proof of Theorem 4.6 evolves completely analogously to the proof of Theorem 4.3 and is left to the reader.

Notice that under the assumptions of Theorem 4.6 Lipschitz continuity results if and only if there exists a real number $\epsilon > 0$ such that for all $T \in S$ and $x_1, x_2 \in W_1$,

$$\|P_T(I + G_2G_1)x_1 - P_T(I + G_2G_1)x\| \geqslant \epsilon \|P_Tx_1 - P_Tx_2\|.$$

It is now a simple matter to use the invertibility theorems of Chapter 2 to obtain specific conditions[16] for continuity. These conditions are stated in the following corollaries.

COROLLARY 4.6.1

The feedback system described by the equations (FE) is Lipschitz continuous if G_1 and G_2 are Lipschitz continuous on their domain and if the open loop is contracting on W_1 (i.e., $\|G_2G_1\|_\Delta < 1$).

COROLLARY 4.6.2

Let $W_1 = W_2$. Then the feedback system described by the equations (FE) is Lipschitz continuous if it is well posed, if G_2 is Lipschitz continuous on W_2, and if there exists a scalar c such that $I + cG_2$ has a causal inverse on W_{1e}, the inverse $(I + cG_1)^{-1}$ is Lipschitz continuous on W_1 and $(G_2 - cI)G_1(I + cG_1)^{-1}$ is contracting on W_1.

COROLLARY 4.6.3

Let $W_1 = W_2$ be a Hilbert space. Then the feedback system described by equations (FE) is Lipschitz continuous if it is well posed, if G_2 is Lipschitz continuous on W_2, and if for some real numbers $a \leqslant b$, $b > 0$, the operator G_2 is incrementally strictly inside the sector $[a,b]$, $I + \frac{1}{2}(a + b)G_1$ has a causal inverse on W_{1e}, and G_1 satisfies one of the following conditions:

1. $a < 0$, and G_1 is incrementally inside the sector $[-1/b, -1/a]$ on W_1;
2. $a > 0$, and G_1 is incrementally outside the sector $[-1/a, -1/b]$ on W_{1e}; or
3. $a = 0$, and $G_1 + (1/b)I$ is incrementally positive on W_{1e}.

[16] These conditions are again essentially due to Zames (Refs. 14, 16). Compare, however, the $\|G_2G_1\|_\Delta < 1$ condition with the $\|G_2\|_\Delta \|G_1\|_\Delta < 1$ condition of Ref. 16.

COROLLARY 4.6.4

Let $W_1 = W_2$ be a Hilbert space. Then the feedback system described
by equations (FE) is Lipschitz continuous if it is well posed, if G_2 is
incrementally positive on W_{2e}, and if G_1 is incrementally strictly
positive and Lipschitz continuous on W_1.

Proofs of these corollaries follow from Theorem 4.6 and Theorem
2.12, Theorem 2.13, Corollary 2.15.1, and Theorem 2.16. The details
are left to the reader.

Corollaries 4.3.1 and 4.3.4 are again very intuitive in nature. In all
of the theorems just cited, the roles of G_1 and G_2 can of course be
interchanged or replaced by a causal factorization of $G_2 G_1$.

4.5.2 Conditions for Discontinuity

It would be possible to give here some specific conditions for discontinuity. These would, however, be identical to the instability
conditions of Theorem 4.4 and Corollary 4.4.1. Since instability implies
discontinuity (stability is actually continuity at the origin), these
conditions are thus quite conservative. Sharper conditions will be given
in Chapter 7 when discussing linearization.

4.6 Concluding Remarks

4.6.1 Sensitivity

A concept related to stability is that of sensitivity. Two types of
sensitivity are usually considered in the literature: the first is sensitivity
with respect to measurement noise, and the second is sensitivity with
respect to modeling errors.

Consider the feedback system shown in Figure 4.6 and the open-loop

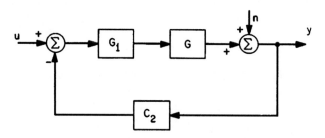

Figure 4.6 Feedback System

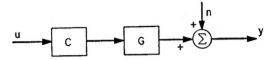

Figure 4.7   Equivalent Open-Loop System

system shown in Figure 4.7. The system G is called the *plant*, C_1 is the *forward-loop compensator*, C_2 is the *feedback-loop compensator*, and C is the *open-loop compensator*. It is assumed that all these operators are well-defined causal mappings between appropriate extended spaces and that they satisfy the usual regularity conditions. Let W_{1e} be the input space and W_{2e} be the output space.

It is assumed that under the nominal operating conditions, $n = 0$, the feedback and the open-loop systems are equivalent. This imposes certain conditions on the compensators C_1, C_2, and C. In fact, this restriction is equivalent to the condition that $GC_1(I + C_2GC_1)^{-1} = GC$ on W_{1e}.

One of the primary purposes of feedback is sensitivity reduction. For output noise this means that for the same noise n, y differs less from its nominal value when the feedback system is used than when the open-loop system is used.

Definition: Let $u \in W_{1e}$ be given and let $y \in W_{2e}$ be the corresponding solution with $n = 0$. This output will of course be the same for the open-loop system as for the feedback system. Let $u \in W_{1e}$ and $n \in W_{2e}$ be given and let $y_1 \in W_{2e}$ be the output of the feedback system and $y_2 \in W_{2e}$ be the output of the open-loop system. Then in general $y_2 \neq y_1$. Let $\delta y_1 = y_1 - y$ and $\delta y_2 = y_2 - y = n$. Then the feedback system is said to *reduce the sensitivity with respect to output noise* if it is well posed and if for all $u_1 \in W_{1e}$, $u_2 \in W_{2e}$, and $T \in S$, $\|P_T \, \delta y_1\| \leqslant \|P_T \, \delta y_2\| = \|P_T n\|$. One can also define in the same way sensitivity reduction along a particular input (usually taken $u = 0$). The particular type of sensitivity reduction one wants again depends on the application.

The relationship between sensitivity reduction and finite gain stability is clear: for sensitivity reduction (around $u = 0$) one wants $\|I + G_2(-G_1)\| \leqslant 1$, whereas for finite gain stability one wants $\|I + G_2(-G_1)\| < \infty$. Stability can thus be regarded as a property of a system with a finite sensitivity coefficient whereas sensitivity reduction can be regarded as a property of a system with a sensitivity coefficient less than or equal to unity.

A second type of sensitivity is with respect to parameter variations and assumes that the plant G is an idealization or an approximation of the actual plant $G + \Delta G$. This thus modifies the outputs, and for some feedback systems this change in outputs is less than the corresponding change for the equivalent open-loop system.

Definition: Let $u \in W_{1e}$ be given and let $y \in W_{2e}$ be the corresponding solution. This output will be the same for the open-loop system as for the feedback system. Let ΔG be a causal operator on the appropriate extended space, and consider on one hand the open-loop system and on the other hand the feedback system with G replaced in both by $G + \Delta G$. This will lead to outputs corresponding to the input $u \in W_{1e}$. Let $y_1 \in W_{2e}$ and $y_2 \in W_{2e}$ be the outputs of, respectively, the feedback and the open-loop systems. Let $\delta y_1 = y_1 - y$ and $\delta y_2 = y_2 - y$. Then the feedback system is said to *reduce the sensitivity with respect to plant variations* if it is well posed and if for all $u \in W_{1e}$, $T \in S$, and operators ΔG such that the feedback system remains well posed, $\|P_T \, \delta y_1\| \leqslant \|P_T \, \delta y_2\|$ for $\|P_T \, \Delta G P_T\|_\Delta$ sufficiently small.[17] This sensitivity reduction can again also be considered for a particular u (usually $u = 0$), and in connection with measurement noise.

No specific conditions for sensitivity reduction will be given. The reader is referred to the specialized literature for results and further discussion.[18] It is the intent of this section to introduce the concept of sensitivity reduction in the framework of extended spaces since no general theory of feedback systems can claim completeness without at least introducing this important concept.

4.6.2 Stability and Continuity for Linear Systems

Linear systems are of particular importance in engineering: they can be analyzed in much more detail than nonlinear systems, and there exist very sophisticated and explicit synthesis techniques for linear systems. It thus comes as no surprise that one resorts to a linear model if this is at all reasonable and that many design techniques are essentially based on the linear system obtained by linearizing a nonlinear model

---

[17] Although it is not clear why only plant variations which are small in the uniform operator topology are discussed in the literature. In view of the discussion in Section 4.3, it appears that plant variations which are small in the strong operator topology would seem to be more appropriate.

[18] The approach taken by Porter (see, for instance, Ref. 19 and further references therein) is similar to the one presented here.

around its desired operating point. The main result of this section states:

THEOREM 4.7 (LINEAR SYSTEMS)

Consider the feedback system described by equations (FE) and assume that it is well posed and that G_1 and G_2 are linear operators respectively on W_{1e} and W_{2e}. Then stability, finite gain stability, continuity, and Lipschitz continuity are equivalent.[19]

Proof: Since the inverse of any linear operator is linear and since boundedness, continuity, and Lipschitz continuity are equivalent for linear operators, the only fact that needs to be proved is that stability implies finite gain stability. Thus, assume that the operator $(I + G)^{-1}$ maps $W_1 \times W_2$ into itself. Let $\{u_n\}$ and $\{(I + G)^{-1}u_n\}$ be convergent sequences in $W_1 \times W_2$. Let $u = \lim_{n \to \infty} u_n$ and

$$e = \lim_{n \to \infty} (I + G)^{-1}u_n,$$

and assume that $e \neq (I + G)^{-1}u$. Then there exists some $T \in S$ such that

$$P_T e \neq P_T(I + G)^{-1}u = P_T(I + G)^{-1}P_T u.$$

Since, however, $P_T(I + G)P_T$ is a bounded operator, its inverse is bounded whenever it exists. This follows from the closed graph theorem (Ref. 20, p. 47). Thus $P_T(I + G)^{-1}P_T$ is bounded and hence continuous which shows that in fact

$$P_T e = \lim_{n \to \infty} P_T e_n = \lim_{n \to \infty} P_T(I + G)^{-1}P_T u_n = P_T(I + G)^{-1}P_T u.$$

This contradiction then establishes that $(I + G)^{-1}$ is a closed operator. Since an everywhere defined closed linear operator is necessarily bounded (see again the closed graph theorem, Ref. 20, p. 47), the theorem follows.

The specific conditions for stability discussed in Section 4.4 can consequently be somewhat simplified for linear systems. The most important generalization which can thus be obtained is a generalization of the small-loop theorem, which states that finite-gain stability results if the spectral radius of $G_2 G_1$, i.e., $\lim_{n \to \infty} \|(G_2 G_1)^n\|^{1/n}$, is less than unity.

---

[19] This result should come as no surprise. Indeed, although it is possible to prove that at least one linear, everywhere-defined, unbounded operator between Banach spaces exists, the existing examples use the axiom of choice in an essential way and are thus not very explicit.

4.6.3 *Doubly Infinite Time-Intervals of Definition*

Besides having dubious physical significance, it turns out that the case in which the time-interval of definition, S, is $(-\infty, +\infty)$, leads to some mathematically very severe questions of well-posedness.

On examination of the definitions of stability and continuity as presented here, one observes that well-posedness of the feedback systems is taken as one of the conditions. This is *not* standard although it is, in the author's opinion, the most logical way to proceed. It is entirely possible, however, to separate the questions of well-posedness on one hand and stability and continuity on the other. The "advantage" of such an approach is that this allows one to study these two questions separately. This, however, is somewhat fallacious since *both* questions will eventually have to be faced and since the most logical approach is to examine the well-posedness question first. Well-posedness indeed expresses a very fundamental property which needs to be satisfied in order for any analysis to be well founded. Moreover, well-posedness and stability are *not really unrelated* in the sense that if a particular feedback system is not well posed, then the model has to be modified, and such a modification will in general alter some fundamental properties of the feedback system including, possibly, its stability properties. It suffices to examine the Nyquist criterion or the circle criterion with the introduction of a small delay in the loop as a convincing illustration of this fact.

If one, however, treats separately the questions of well-posedness and stability or continuity, then it turns out that the specific conditions obtained in Corollaries 4.3.1, 4.3.2, 4.3.3, and 4.3.4 and Corollaries 4.5.1, 4.5.2, 4.5.3, and 4.5.4 are in fact still valid as conditions for stability or continuity.

As far as Theorem 4.4 on instability is concerned, there is then of course no need for backwards extensions of the involved operators when $S = (-\infty, +\infty)$ and instability results with the conditions of the theorem or its corollary on the operators G_1 and G_2 directly. The basic idea behind the resulting instability theorem is essentially the following: Let $S = (-\infty, +\infty)$. Then the feedback system described by equations (FE) is discontinuous (which is then defined *without* involving well-posedness) if the operator $(I + G_2G_1)$ has a noncausal inverse on W_1. This in fact then leads to the following phenomenon: Assume that a feedback system is discontinuous with $S = (-\infty, +\infty)$ and that $I + G_2G_1$ is invertible on W_1 but that this inverse is not causal (not all discontinuous systems need have this property but some do, as illustrated by Theorem 4.4). Then $I + G_2G_1$ is *not* invertible on $W_{1T} \triangleq \{x \in W \mid P_Tx = 0, \ T \in (-\infty, +\infty) \text{ given}\}$ for *at least one* $T \in (-\infty, +\infty)$. Weak additional assumptions on G_2G_1 (see Lemma 2.2) thus imply that in fact $(I + G_2G_1)$ will not be invertible on $W_{1T'} \triangleq \{x \in W \mid P_{T'}x = 0, \ T' \in (-\infty, +\infty) \text{ given}\}$ for *any* $T' \in (-\infty, +\infty)$. Again, weak assumptions on G_2G_1 (see Section 4.3) ensure that $(I + G_2G_1)$ will on the other hand have a causal inverse on $W_{1T'e}$ for any $T' \in (-\infty, +\infty)$. Hence for such discontinuous systems there exists for *any* $T' \in (-\infty, +\infty)$, an input $u_1 \in W_1$ and $u_2 = 0$ with $P_{T'}u_1 = 0$ such that at least two solutions exist to the equations describing the feedback system (and thus for linear systems an infinite number of solutions exist). One solution yields $e_1 \in W$ through $e_1 = (I + G_2G_1)^{-1}u_1$ with $(I + G_2G_1)^{-1}$ the noncausal inverse on W_1, and the other yields $e_1 \in W_{1T'e}$, $e_1 \notin W_1$, through $e_1 = (I + G_2G_1)^{-1}u_1$, with $(I + G_2G_1)^{-1}$ the causal inverse on $W_{1T'e}$. It is clear that the latter one is the "physical" solution in the dynamical-system sense and that the first one (which extends to $-\infty$) is a mathematical artifact.

A study of the above issues leads to at least one partial result concerning well-posedness of feedback systems with $S = (-\infty, +\infty)$. Since it is of some mathematical interest, it is given below. First, however, one more definition is given.

Definition: Let $S = (-\infty, +\infty)$, let W be a Banach space, and W_e be its extension. Assume that the family of projection operators $\{P_T\}$, $T \in S$, is a resolution of the identity on W; i.e., for all $x \in W$, $\lim_{T \to -\infty} P_Tx = 0$ and $\lim_{T \to \infty} P_Tx = x$ and let G be a causal operator from W_e into itself. Then the feedback system described by the equation $(I + G)e = u$ is said to be *continuous* if for any given $u \in W$ every solution $e \in W_e$ to the above equation actually yields $e \in W$ and if given any $u \in W$ and $\epsilon > 0$ there exists a $\delta > 0$ such that if $\|\delta u\| \leqslant \epsilon$ then any solutions e and $e + \delta e$ corresponding to, respectively, u and $u + \delta u$ satisfy $\|\delta e\| \leqslant \delta$.

THEOREM 4.8

Assume that G is a causal operator on W_e and that $(I + G)$ has, for any $T \in S = (-\infty, +\infty)$, a causal inverse on $W_{Te} \triangleq \{x \in W_e \mid P_Tx = 0, \ T \in S \text{ given}\}$. Then the feedback system described by the equation $(I + G)e = u$ is continuous if and only if $I + G$ has a continuous causal inverse on W (and hence on W_e).

Proof: Since the feedback system is continuous and since $I + G$ has, for any $T \in S$, a causal inverse on W_{Te}, $(I + G)^{-1}$ maps for any $T \in S$,

$W_T \triangleq \{x \in W \mid P_T x = 0\}$ into itself and is continuous on W. Since $\{P_T\}$, $T \in S$, is a resolution of the identity, the family of subspaces $\{W_T\}$, $T \in S$, is dense in W. Since $(I + G)^{-1}$ is thus continuous on a dense set, a continuous inverse exists on the whole space, and this inverse may be obtained by a continuous extension of $(I + G)^{-1}$. A simple argument by contradiction then shows that this inverse is causal as well.

The value of the Theorem 4.8 appears to be that it gives a condition for well-posedness (causal invertibility of $(I + G)$ on W_e — other considerations are ignored here) when the time interval of definition S is $(-\infty, +\infty)$. This condition is very strong indeed, since it requires continuity of the feedback system.

References

1. Zames, G., "Realizability Conditions for Nonlinear Feedback Systems," *IEEE Trans. on Circuit Theory*, Vol. CT-11, pp. 186–194, 1964.
2. Chu, S. C., and Diaz, J. B., "On 'in the Large' Application of the Contraction Principle," in *Differential Equations and Dynamical Systems* (edited by LaSalle and Hale), pp. 235–238, Academic Press, New York, 1967.
3. Moyer, R. D., "Uniqueness Theorems for Initial Value Problems," in *Differential Equations and Dynamical Systems* (edited by LaSalle and Hale), pp. 505–509, Academic Press, New York, 1967.
4. Saaty, T. L., *Modern Nonlinear Equations*, McGraw-Hill, New York, 1967.
5. Willems, J. C., "On the Well-Posedness of Feedback Systems," to appear.
6. Willems, J. C., "Stability, Instability, Invertibility and Causality," *SIAM J. on Control*, Vol. 7, No. 4, pp. 645–671, 1969.
7. Nyquist, H., "Regeneration Theory," *Bell System Tech. J.*, Vol. 2, pp. 126–147, 1932.
8. Brockett, R. W., "The Status of Stability Theory for Deterministic Systems," *IEEE Trans. on Automatic Control*, Vol. AC-11, pp. 596–606, 1966.
9. Brockett, R. W., *Finite Dimensional Linear Systems*, John Wiley and Sons, New York, 1970.
10. Popov, V. M., "Absolute Stability of Nonlinear Systems of Automatic Control," *Automation and Remote Control*, Vol. 22, pp. 961–979, 1961.
11. Kalman, R. E., "Liapunov Functions for the Problem of Luré in Automatic Control," *Proc. Natl. Acad. Sci.*, Vol. 49, pp. 201–205, 1963.
12. Willems, J. C., "A Survey of Stability of Distributed Parameter Systems," in *Control of Distributed Parameter Systems*, pp. 63–102, ASME Publications, 1969.
13. Sandberg, I. W., "On the L_2-Boundedness of Solutions of Nonlinear Functional Equations," *Bell System Tech. J.*, Vol. 43, pp. 1581–1599, 1964.

14. Zames, G., "On the Stability of Nonlinear, Time-Varying Feedback Systems," *Proc. 1964 Natl. El. Conf.*, Vol. 20, pp. 725–730, 1964.
15. Sandberg, I. W., "Some Results on the Theory of Physical Systems Governed by Nonlinear Functional Equations," *Bell System Tech. J.*, Vol. 44, pp. 871–898, 1965.
16. Zames, G., "On the Input-Output Stability of Time-Varying Nonlinear Feedback Systems. Part I: Conditions Derived Using Concepts of Loop Gain, Conicity, and Positivity. Part II: Conditions Involving Circles in the Frequency Plane and Sector Nonlinearities," *IEEE Trans. on Automatic Control*, Vol. AC-11, pp. 228–238 and 465–476, 1966.
17. Brockett, R. W., and Willems, J. L., "Frequency Domain Stability Criteria: Part I and II." *IEEE Trans. on Automatic Control*, Vol. AC-10, pp. 255–261 and 401–413, 1965.
18. Brockett, R. W., and Lee, H. B., "Frequency Domain Instability Criteria for Time-Varying and Nonlinear Systems," *IEEE Proceedings*, Vol. 55, pp. 604–619, 1967.
19. Porter, W. A., and Zahm, C. L., *Basic Concepts in System Theory*, SEL Technical Report 44, University of Michigan, Ann Arbor, 1969.
20. Hille, E., and Phillips, R. S., *Functional Analysis and Semi-Groups* (second edition), American Mathematical Society, Providence, 1957.

5 The Nyquist Criterion and the Circle Criterion

The results obtained in the previous chapter will now be applied to a particular class of linear feedback systems. They consist of a linear time-invariant system in the forward loop and a linear memoryless gain in the feedback loop. The case in which this gain is constant will be given special attention, and for this case necessary and sufficient conditions for stability will be obtained. For the time-varying case sufficient conditions for stability and instability will be derived. The results presented in this chapter are extensions of the Nyquist criterion and the circle criterion.

5.1 Mathematical Description of the Feedback System

Consider the feedback loop shown in Figure 5.1. In terms of the notation used in the previous chapter let the time-interval of definition $S = [T_0, \infty)$, $V_1 = V_2 = R$, i.e., u_1, e_1, y_1, u_2, e_2, y_2 are real-valued functions of time on $[T_0, \infty)$, and let G_1 and G_2 be formally defined by

$$(G_1 x)(t) \triangleq \sum_{n=0}^{\infty} g_n x(t - t_n) + \int_{T_0}^{\infty} g(t - \tau) x(\tau) \, dt$$

$$(G_2 x)(t) \triangleq k(t) x(t)$$

where $\{t_n\}$, $n \in I^+$, is a sequence of real numbers with $t_0 = 0$, $t_n > 0$ for $n \geqslant 1$, $g(t)$ is a real-valued function on R with $g(t) = 0$ for $t < 0$ and

$g \in L_{1e}(0,\infty)$, $\{g_n\}$, $n \in I^+$, is a real-valued sequence satisfying

$$\sum_{\{n \mid t_n \leqslant T\}} |g_n| < \infty$$

for all $T \in [0,\infty)$, and $k(t)$ is a real-valued function on $[T_0,\infty)$ with $k(t) \in L_{\infty e}(T_0,\infty)$.

The functional equations describing the feedback system are

$$e_1(t) = u_1(t) - y_2(t)$$

$$e_2(t) = u_2(t) + y_1(t)$$

$$y_1(t) = (G_1 e_1)(t) \tag{LFE}$$

$$y_2(t) = (G_2 e_2)(t)$$

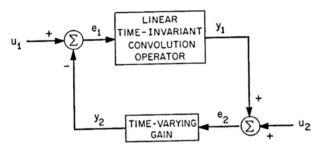

Figure 5.1   The Linear Feedback System under Consideration

The solution space will be taken as $L_{pe}(T_0,\infty)$, $1 \leqslant p \leqslant \infty$, with $L_{pe}(T_0,\infty)$ defined in the usual way. The inputs u_1, u_2 are assumed to be given elements of $L_{pe}(T_0,\infty)$.

A solution to the above feedback equations thus consists of a quadruple of real-valued functions on $[T_0,\infty)$, $\{e_1,e_2,y_1,y_2\}$ with e_1, e_2, y_1, $y_2 \in L_{pe}(T_0,\infty)$ and which satisfy the feedback equations (LFE) for almost all $t \geqslant T_0$.

The operators G_1 and G_2 have been studied in detail in Chapter 2. It follows readily from Minkowski's inequality and some direct estimates that G_1 and G_2 map $L_{pe}(T_0,\infty)$ into itself. They are in fact both locally Lipschitz-continuous linear operators on $L_{pe}(T_0,\infty)$, but for boundedness additional conditions are necessary. G_1 is moreover a time-invariant operator and G_2 is time invariant if and only if $k(t)$ is almost everywhere equal to a constant.

The feedback system described by the equations (LFE) thus satisfies the assumptions W.1–W.4, G.1–G.3, and I.1, and the theory developed in Chapter 4 is applicable.

The values of the gain $k(t)$ is of course what is important for the characterization of the operator G_2. The operator G_1 has been characterized by the function g and the sequence $\{g_n, t_n\}$, but it is, both from a mathematical and an engineering point of view, much more conveniently characterized by its Laplace transform. More precisely, assume that for some real number σ, $g(t)e^{-\sigma t} \in L_1(0,\infty)$ and $\{g_n e^{-\sigma t_n}\} \in l_1$. Then the *Laplace transform of G_1*,

$$G(s) \triangleq \sum_{n \in I^+} g_n e^{-st_n} + \int_0^\infty g(t)e^{-st}\, dt$$

exists for $\mathrm{Re}\, s \geqslant \sigma$ and is analytic in $\mathrm{Re}\, s > \sigma$. Let $x(t)$ be such that for some real number σ', $x(t)e^{-\sigma' t} \in L_1(T_0,\infty)$ then

$$X(s) \triangleq \int_{T_0}^\infty x(t)e^{-st}\, dt$$

exists for $\mathrm{Re}\, s \geqslant \sigma'$ and the function

$$Y(s) \triangleq \int_{T_0}^\infty (G_1 x)(t)e^{-st}\, dt$$

exists for $\mathrm{Re}\, s \geqslant \sigma, \sigma'$ and in fact equals $G(s)X(s)$.

5.2 Well-Posedness

The first question is that of well-posedness. A simple additional condition is therefore required. More precisely:

THEOREM 5.1

Consider the feedback system described by the functional equations (LFE) with $L_{pe}(T_0,\infty)$, $1 \leqslant p \leqslant \infty$, as the solution space. Then this feedback system is well posed if and only if $\|g_0 k(t)\|_{L_\infty(T_0,\infty)} < 1$.

Proof: The sufficiency part of the theorem is a direct application of Corollary 4.1.3. Indeed, the open-loop operator is the sum of a strongly causal operator and a feedthrough which is precisely the time-varying gain $g_0 k(t)$. Since the gain on $L_p(T_0,T)$, $1 \leqslant p \leqslant \infty$, is given by $\|g_0 k(t)\|_{L_\infty(T_0,T)} \leqslant \|g_0 k(t)\|_{L_\infty(T_0,\infty)}$, sufficiency follows.

Next, assume that the gain $k(t)$ is replaced by the gain $k'(t) = k(t) + r$. Then for some $|r| \leqslant \epsilon$, no matter how small ϵ, $|g_0 k'(t)| > 1$

on a set on nonzero measure. Assume now that a delay of length $\epsilon' \neq 0$ is inserted in the loop and consider as the input a unit step starting at $T'' \geqslant T_0$, where T'' is such that any interval $[T'', T'' + \sigma]$, $\sigma > 0$, contains a set of nonzero measure, where $|g_0 k'(t)| \geqslant 1$. It can then be shown that the response to this input will lead to an ill-posed situation on $L_{pe}(T_0, \infty)$, $1 \leqslant p \leqslant \infty$, in the sense that $\lim_{\epsilon' \to 0} \lim_{\epsilon \to 0} e_{\epsilon' \epsilon}$ will not equal the response obtained by taking $\epsilon' = \epsilon = 0$. The mechanics of this explicit calculation are similar to those used in Corollary 4.1.3. The details are left to the reader.

Theorem 5.1 shows, among other things, that the case where the forward loop represents a strongly causal system ($g_0 = 0$) is the only one which should be used in connection with an unbounded gain $k(t)$ in the feedback loop.

5.3 Stability and Instability in the Time-Invariant Case

The stability results obtained in this section essentially constitute the classical Nyquist criterion.[1] The case under consideration is where the feedback gain $k(t)$ equals a constant almost everywhere and where the system is open-loop stable (the kernel of the forward-loop convolution operator is integrable). First, however, several definitions are introduced.

Definitions: Let LA denote the algebra consisting of elements determined by a real-valued $L_1(-\infty, +\infty)$-function g, a real-valued l_1-sequence $\{g_n\}$, $n \in I^+$, and a sequence of real numbers $\{t_n\}$, $n \in I^+$. Addition of $x_1 = (g_1, \{g_n, t_n\}_1)$ and $x_2 = (g_2, \{g_n, t_n\}_2)$ is defined as $x_1 + x_2 \triangleq (g_1 + g_2, \{g_n, t_n\}_3)$, where the sequence $\{g_n, t_n\}_3$ consists of exactly all pairs $\{g_n, t_n\}_1$ and $\{g_n, t_n\}_2$. Multiplication by scalars is defined as $\alpha x = \alpha(g, \{g_n, t_n\}) \triangleq (\alpha g, \{\alpha g_n, t_n\})$, and multiplication of elements is

---

[1] The Nyquist criterion is well known, although many of the engineering texts which treat this stability problem *a priori* equate stability with the absence of right-half-plane singularities of the closed-loop frequency response. Although this is clearly valid for instance in the case of rational functions, this equivalence is far from obvious for more general transfer functions. The results stated in these texts should thus be carefully interpreted when applied to nonrational transfer functions since it is known that one indeed can get into difficulties for sufficiently complex transfer functions (see Ref. 1). Recently Desoer (Refs. 1-4), recognizing this difficulty, has obtained rigorous derivations to essentially the level of generality treated in this chapter (but with multiple inputs). These papers, however, treat only stability, whereas the results obtained in this section treat instability as well.

defined by

$$x_1 x_2 \triangleq \left(\int_{-\infty}^{+\infty} g_1(t - \tau) g_2(\tau) \, d\tau + \sum_n (g_n)_1 g_2(t - (t_n)_1) \right.$$

$$\left. + \sum_n (g_n)_2 g_1(t - (t_n)_2), \{g_n, t_n\}_3 \right),$$

where $\{t_n\}_3 = \{t_n\}_1 \oplus \{t_n\}_2$ (i.e., all elements of the form $t_n = t_{n_1} + t_{n_2}$ where t_{n_1} and t_{n_2} range over $\{t_n\}_1$ and $\{t_n\}_2$ respectively), and the element g_n corresponding to t_n in $\{g_n, t_n\}_3$ is given by $g_{n_1} g_{n_2}$, with $t_n = t_{n_1} + t_{n_2}$. Let the norm on LA be defined by $\|g\|_{L_1} + \|\{g_n\}\|_{l_1}$. It can be shown that LA as defined above is a real commutative Banach algebra with the unit $e \triangleq (0, \{g_n, t_n\})$ where $g_0 = 1$, $t_0 = 0$ and $g_n = 0$ otherwise.

Consider now the subset of LA, LA^+, which consist of all elements of LA which satisfy $g(t) = 0$ for $t < 0$ and $t_n \geqslant 0$ for all n. It is easily verified that LA^+ is a subalgebra and that it contains the unit. The details of the proofs of the above claims can be found elsewhere (Ref. 5, pp. 141–157).

The *Laplace transform* of an element of LA is defined as the function of the complex variable s defined by

$$G(s) \triangleq \sum_{n=0}^{\infty} g_n e^{-st_n} + \int_{-\infty}^{\infty} g(t) e^{-st_n} \, dt.$$

It is well defined and uniformly continuous along $\text{Re } s = 0$ for elements of LA. It is well defined and analytic in $\text{Re } s > 0$ for elements of $(LA)^+$.

The importance of the algebras LA and LA^+ stems from the fact that they define subspaces of $\mathscr{L}(L_p(-\infty, +\infty), L_p(-\infty, +\infty))$, $1 \leqslant p \leqslant \infty$: namely those subspaces formed by convolution operators with kernels consisting of an integrable function and a summable string of impulses. In fact, LA represents "almost" all time-invariant operators in $\mathscr{L}(L_p, L_p)$ and by and large all those of interest in mathematical systems theory. Multiplication and addition in LA correspond to those operations in $\mathscr{L}(L_p, L_p)$. Moreover, LA forms a regular subalgebra of $\mathscr{L}(L_1, L_1)$, $\mathscr{L}(L_2, L_2)$, and $\mathscr{L}(L_\infty, L_\infty)$ in the sense that every element of LA which is invertible in $\mathscr{L}(L_2, L_2)$, for example, is invertible in LA itself. LA^+ is of particular importance; the corresponding elements of $\mathscr{L}(L_p, L_p)$, $1 \leqslant p \leqslant \infty$, are causal.

The next lemma plays an essential role in the results which follow.

LEMMA 5.1

Let $(g,\{g_n,t_n\}) \in LA^+$. Then $(g,\{g_n,t_n\})$ is invertible (regular) in LA^+ if and only if $\inf_{\text{Re } s \geqslant 0} |G(s)| > 0$, and in LA if and only if $\inf_{\text{Re } s = 0} |G(s)| > 0$.

A proof of this classical result can be found in most books treating convolution operators (Ref. 5, pp. 150, 155).

Application of the above lemma leads to the following necessary and sufficient condition for stability of linear time-invariant feedback systems.

THEOREM 5.2

Assume that the feedback system described by the functional equations (LFE) is well-posed and that $k(t) = K$ almost everywhere. Assume furthermore that $\{g_n\} \in l_1$ and $g \in L_1(0,\infty)$ (i.e., the feedback system is open-loop stable). Let $G(s)$ denote the Laplace transform of $(g,\{g_n,t_n\})$. Let $L_2(T_0,\infty)$ be the space with respect to which stability is defined. Then the feedback system is stable if and only if

$$\inf_{\text{Re } s \geqslant 0} |1 + KG(s)| > 0.$$

Proof: It suffices by Theorem 4.3 to demonstrate that the operator $I + G_2'G_1$ is invertible on $L_2(T_0,\infty)$ if and only if the conditions of the Theorem are satisfied. There are three mutually exclusive possibilities:

1. $\inf_{\text{Re } s \geqslant 0} |1 + KG_1(s)| > 0$;

2. $\inf_{\text{Re } s = 0} |1 + KG_1(s)| = 0$; or

3. $\inf_{\text{Re } s > 0} |1 + KG_1(s)| = 0$ and $\inf_{\text{Re } s = 0} |1 + KG_1(s)| > 0$.

It needs to be shown that case 1 yields invertibility and that in cases 2 and 3, the operator $I + G_2G_1$ is not invertible on $L_2(T_0,\infty)$. In the first case, it follows from Lemma 5.1 that $I + G_1G_2$ has a bounded causal inverse on $L_2(T_0,\infty)$ which yields stability. Assume next that case 2 is satisfied; then $I + G_2G_1$ multiplies the limit-in-the-mean transform of the element on which it operates by $1 + KG(j\omega)$ and thus the only candidate for the inverse is the operator which divides the limit-in-the-mean transform of the element on which it operates by $1 + KG(j\omega)$. Thus for this inverse to be bounded, $(1 + KG_1(j\omega))^{-1}$ ought to exist for almost all $\omega \in R$ and belong to L_∞. Since $G_1(j\omega)$ is continuous and, by assumption, $\inf_{\omega \in R} |1 + KG(j\omega)| = 0$, the operator $I + G_2G_1$ has in that case no bounded inverse on $L_2(T_0,\infty)$, which thus yields instability. Assume finally that case 3 is satisfied. Lemma 5.1 then implies that the operator $I + G_2'G_1'$, defined on $L_2(-\infty,+\infty)$ by the same formal expressions as G_1 and G_2 (but with the lower limit in the convolution integral defining G_2 replaced by $-\infty$), has a noncausal inverse on $L_2(-\infty,+\infty)$, which thus yields the noninvertibility of $I + G_2G_1$ by an argument similar to the one used to prove Theorem 2.22.

Remark 1: The above theorem is well known, although the usual proofs assume the equivalence of stability and the absence of singularities of $(1 + KG(s))^{-1}$ in Re $s \geqslant 0$, and lack therefore a certain amount of justification (see Ref. 1). Notice that since the system is linear, instability implies that there exists an input $u_1 \in L_2(T_0,\infty)$ with $u_2 = 0$ such that e_1 and $e_2 \in L_{2e}(T_0,\infty) - L_2(T_0,\infty)$.

Remark 2: If stability is defined with respect to $L_p(T_0,\infty)$, $1 \leqslant p \leqslant \infty$, then it is clear that the condition of the theorem is still sufficient for stability. If the condition fails because case 3 prevails, then instability results. These claims follow the proof of Theorem 5.2. If the condition fails because case 2 prevails, the situation is more complex. It can still be shown that then instability results at least when $p = 1$ or ∞, since $I + G_2G_1$ is invertible on $L_p(T_0,\infty)$, $p = 1$, ∞, *if and only if* it is invertible in LA^+.

Remark 3: It is possible to verify, at least in some cases, the condition $\inf_{\text{Re}\,s\geqslant 0}|1 + KG_1(s)| > 0$ by: 1, establishing that $\inf_{\omega \in R}|1 + KG(j\omega)| > 0$; and 2, checking whether $KG(j\omega)$ encircles the $-1 + 0j$ point.[2] It has not been possible as yet to completely generalize this condition to the case under consideration, mainly because it appears to be no easy matter to give a suitable generalization of the no-encirclement condition. One important particular case is stated below, namely, when the delays are equally spaced, i.e., when $t_n = nT$ for some $T > 0$.

Definition: The *argument* of $1 + KG(j\omega)$, denoted by $\theta(\omega)$, with $1 + KG(j\omega) \neq 0$, $\omega \in R$, is defined as the *continuous* function with $\theta(0) = 0$ such that for all $\omega \in R$, $1 + KG(j\omega) = |1 + KG_1(j\omega)|e^{j\theta(\omega)}$

THEOREM 5.3

The condition $\inf_{\text{Re}\,s\geqslant 0}|1 + KG(s)| > 0$ is equivalent to the following conditions:

1. $\inf_{\text{Re}\,s=0}|1 + KG(s)| > 0$, and
2. $\lim_{N\to\infty}\theta(N2\pi T^{-1})$ exists and is zero.

[2] For a proof of this case when $g_n = 0$ for all $n \neq 0$ see Reference 6.

Proof: Let $A(s) \triangleq \sum_{n=0}^{\infty} g_n e^{-snT}$ and $L(s) \triangleq \int_0^{\infty} g(t)e^{-st}\, dt$. The function $G(j\omega) = A(j\omega) + L(j\omega)$ is the sum of a periodic function $A(j\omega)$, and a bounded function $L(j\omega)$ that, by the Riemann-Lebesque lemma (Ref. 7, p. 103), approaches zero as $|\omega| \to \infty$. Since $\inf_{\text{Re}\,s=0} |1 + KG(s)| > 0$ by condition 1, it follows that $\inf_{\omega \in R} |1 + KA(j\omega)| > 0$. Since $\lim_{N \to \infty} \theta(N2\pi T^{-1})$ *exists* by condition 2, it follows that the argument $\Phi(\omega)$ of $1 + KA(j\omega)$ satisfies $\Phi(2\pi T^{-1}) = \Phi(0)$. Thus by the principle of the argument (Ref. 7, p. 216), there are no zeros of the function $R(z) = \sum_{n=0}^{\infty} g_n z^n$ inside the unit circle since $R(z)$ is analytic inside the unit circle and since the increase in its argument as z moves around the unit circle equals zero. Thus the function $1 + KA(s)$ has no zeros in $\text{Re}\,s \geqslant 0$. Consider now the contour in the complex plane shown in Figure 5.2. The increase of the argument of $1 + KG(s)$ as s

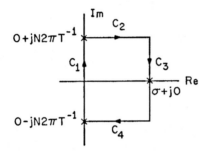

Figure 5.2 A contour in the Complex Plane

moves around this contour is zero for N and σ sufficiently large. Indeed, along C_1 it is zero by the assumption $\lim_{N \to \infty} \theta(N2\pi T^{-1}) = 0$, and along C_2, C_3, C_4 it is zero since $G(s)$ is arbitrarily close to $A(s)$ along that part of the contour. Hence, $1 + KG(s)$ has by the principle of the argument no zeros in any finite part of the half-plane $\text{Re}\,s \geqslant 0$. It is bounded away from zero in $\text{Re}\,s \geqslant 0$ since it arbitrarily closely approximates $|1 + KG(s)|$ for large values of $|s|$ in $\text{Re}\,s \geqslant 0$. Thus, $|1 + KG(s)|$ is indeed bounded away from zero in $\text{Re}\,s \geqslant 0$, as claimed. This argument of this proof is easily reversed to yield the converse of the theorem.

Theorem 5.3 yields a systematic procedure for obtaining the ranges of the feedback gain K which yield stability and instability. This is illustrated in Figure 5.3 and requires verifying: 1, that the $-1/K$ point

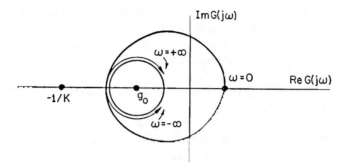

Figure 5.3 Illustrating the Nyquist Criterion

does not belong to the Nyquist locus $G(j\omega)$; and 2, that the Nyquist locus does not encircle the $-1/K$ point.

5.4 Stability and Instability in the Time-Varying Case

All elements are now available to state the circle criterion[3] as applied to linear time-varying feedback systems. Let $k_{min} < k_{max}$ be real numbers with $k_{max} > 0$, and let the *critical disk* $\mathscr{D}$, be defined as

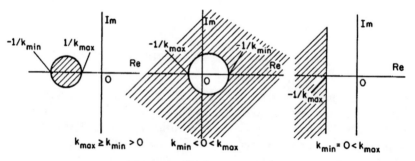

Figure 5.4 The Critical Disk

[3] There are several more or less independent and simultaneous sources for the stability part of the circle criterion. The work of Sandberg (Refs. 8, 9) and Zames (Refs. 6, 10) has influenced attempts to obtain these results in an input-output stability setting. Other treatments of the circle criterion will be found in References 11, 12. The instability part of the circle criterion originates — in a more restricted setting — in Brockett and Lee (Ref. 13), where it is obtained using Lyapunov methods. It was obtained by the author (Ref. 14) in the framework of input-output stability. The circle criterion as it is presented here is in essence a direct generalization of the Nyquist criterion and should thus be of considerable interest in engineering design. Note that the open loop unstable case is not treated here, in contradistinction to Reference 13.

follows (see Figure 5.4):

1. If $k_{max} \geqslant k_{min} > 0$, then $\mathscr{D}$ denotes the closed disk centered on the negative real axis of the complex plane which passes through the points $-1/k_{min}$ and $-1/k_{max}$.
2. If $k_{min} < 0$ and $k_{max} > 0$, then $\mathscr{D}$ denotes the outside of the open disk centered on the negative real axis of the complex plane which passes through the points $-1/k_{min}$ and $-1/k_{max}$.
3. If $k_{min} = 0$ and $k_{max} > 0$, then $\mathscr{D}$ denotes the closed half-plane $\text{Re } s \leqslant -1/k_{max}$.

Let $G(s)$ be the Laplace transform of $(g,\{g_n,t_n\})$ and let $\mathscr{G}^+$ denote the set in the complex plane determined by $\mathscr{G}^+ \triangleq \{G(s) \mid \text{Re } s \geqslant 0\}$ and let $\mathscr{G}^0$ denote the set in the complex plane determined by

$$\mathscr{G}^0 \triangleq \{G(s) \mid \text{Re } s = 0\}.$$

Let

$$d(\mathscr{D},\mathscr{G}^+) = \inf_{\substack{x \in \mathscr{D} \\ y \in \mathscr{G}^+}} |x - y|,$$

and let $d(\mathscr{D},\mathscr{G}^0)$ be similarly defined. The set $\mathscr{G}^0$ is sometimes called the *Nyquist locus* of G_1.

THEOREM 5.4

Assume that $\{g_n\} \in l_1$ and $g \in L_1(0,\infty)$ in the feedback system described by the functional equations (LFE), i.e., that the feedback system is open-loop stable. Let $G(s)$ denote the Laplace transform of $(g,\{g_n,t_n\})$. Let $L_2(T_0,\infty)$ be the space with respect to which stability is defined and assume that the feedback system is well posed. Then the feedback system is stable if there exist real numbers $k_{min} \neq k_{max}$ with[4] $k_{max} > 0$ such that $k_{min} \leqslant k(t) \leqslant k_{max}$ for almost all $t \geqslant T_0$ and if $d(\mathscr{D},\mathscr{G}^+) > 0$.

Proof: It suffices to verify that the assumptions of the theorem imply that the conicity conditions of Corollary 4.3.3 are verified. Assume first that k_{min} and k_{max} are nonzero. Let $k_{min'} < k_{min}$ and $k_{max'} > k_{max}$ be real numbers such that the disk $\mathscr{D}_1$ determined by $k_{min'}$ and $k_{max'}$ still satisfies the condition $d(\mathscr{D}_1,\mathscr{G}^+) > 0$. Such numbers $k_{max'}$ and $k_{min'}$ exist since $\mathscr{D}_1$ is arbitrarily close to $\mathscr{D}$ for $k_{max'} - k_{max}$ and $k_{min} - k_{min'}$ sufficiently small but positive. It is now easily verified that G_2 is strictly inside the sector $[k_{min'},k_{max'}]$ on $L_2(T_0,\infty)$ since $k_{min'} < k_{min} \leqslant k(t) \leqslant k_{max} < k_{max'}$. Furthermore since $|g_0(k_{min'} + k_{max'})/2| < 1$, $I + G_1(k_{min'} + k_{max'})/2$ has by Lemma 5.1 a causal inverse on

---

[4] Clearly assuming $k_{max} > 0$ does *not* constitute any loss of generality since this can always be achieved by replacing G_2 by $-G_2$ and G_1 by $-G_1$.

$L_{2e}(T_0,\infty)$. The only thing that remains to be shown is that G_1 satisfies the appropriate conicity conditions. Assume first that $k_{min} < 0$. It is then clear that G_1 is inside the sector $[-1/k_{max'}, -1/k_{min'}]$ on $L_2(T_0,\infty)$ since G_1 corresponds to multiplication of the limit-in-the-mean transform by $G(j\omega)$ and $d(\mathscr{D}_1,\mathscr{G}^0) > 0$. Assume next that $k_{min} > 0$. It then suffices to prove that G_1 is outside the sector $[-1/k_{min'}, -1/k_{max'}]$ on $L_{2e}(T_0,\infty)$. Since $I + G_1(k_{min'} + k_{max'})/2$ has a bounded causal inverse on $L_2(T_0,\infty)$ by Lemma 5.1, it thus suffices to show (again by considering limit-in-the-mean transforms) that

$$\|(1 + \tfrac{1}{2}(k_{min'} + k_{max'})G(j\omega))^{-1}G(j\omega)\|_{L_\infty} \leqslant 2(k_{max'} - k_{min'})^{-1},$$

which is indeed implied by the frequency domain condition of the theorem. The only case remaining is when $k_{min} = 0$. This case, however, follows from positivity considerations.

Remark: If $k(t)$ has no specified upper bound but if $k(t) \geqslant k_{min}$, then stability results if one of the following conditions is satisfied:

1. $k_{min} > 0$ and $d(\mathscr{D},\mathscr{G}^+) > 0$, where $\mathscr{D}$ denotes the closed disk centered on the negative real axis which passes through the origin and $-1/k_{min}$;
2. $k_{min} < 0$ and $d(\mathscr{D},\mathscr{G}^+) > 0$, where $\mathscr{D}$ denotes the outside of the open disk centered on the negative real axis which passes through the origin and the $-1/k_{min}$ points; or
3. $k_{min} = 0$ and Re $G(s) \geqslant \epsilon > 0$ for all Re $s \geqslant 0$.

In theory it is not necessary for the third condition that $k(t)$ be bounded, provided it belongs to $L_{\infty e}(T_0,\infty)$.

THEOREM 5.5

Assume that $\{g_n\} \in l_1$ and $g \in L_1(0,\infty)$ in the feedback system described by the functional equations (LFE), i.e., that the feedback system is open-loop stable. Let $G(s)$ denote the Laplace transform of $\{g,\{g_n,t_n\}\}$. Let $L_2(T_0,\infty)$ be the space with respect to which stability is defined and assume that the feedback system is well-posed. Then the feedback system is unstable if there exist real numbers $k_{min} \neq k_{max}$ such that $0 < k_{min} \leqslant k(t) \leqslant k_{max}$ for almost all $t \geqslant t_0$ and if $d(\mathscr{D},\mathscr{G}^0) > 0$ and $d(\mathscr{D},\mathscr{G}^+) = 0$.

Proof: Let G_1' and G_2' denote the operators defined on $L_2(-\infty,+\infty)$ by

$$(G_1'x)(t) = \sum_{n=0}^{\infty} g_n x(t - t_n) + \int_{-\infty}^{\infty} g(t - \tau)x(\tau)\,d\tau$$

and

$$(G_2'x)(t) = k'(t)x(t)$$

with

$$k'(t) = \begin{cases} \frac{1}{2}(k_{min} + k_{max}) & \text{for} \quad t < T_0 \\ k(t) & \text{otherwise.} \end{cases}$$

It is a simple matter to verify that $L_2(-\infty, +\infty)$, G_1', and G_2' qualify as backward extensions of $L_2(T_0, \infty)$, G_1, and G_2 and that all the conditions of Corollary 4.4.1 are satisfied. The estimates involved in this verification are in fact identical to the ones used in Theorem 5.4, and the noncausality follows from Lemma 5.1.

Remark 1: It is again possible at least in the case where all the delays are equally spaced, i.e., $t_n = nT$, $T > 0$, to rephrase the conditions of Theorems 5.4 and 5.5 exclusively in terms of the frequency response of the forward loop. In fact $d(\mathscr{D}, \mathscr{G}^+) > 0$ if and only if: 1, $d(\mathscr{D}, \mathscr{G}^0) > 0$; and 2, $\lim_{N \to \infty} \Theta(N2\pi T^{-1})$ exists and is zero, where Θ is the argument of $1 + \alpha G(j\omega)$ and α is an arbitrary element of $\mathscr{D}$. This leads to the situation shown in Figure 5.5.

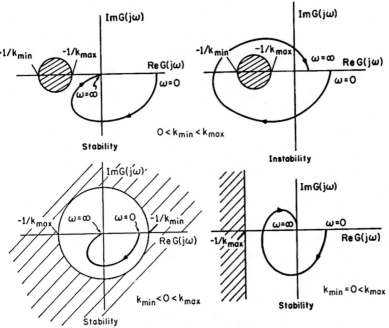

Figure 5.5   Illustrating the Circle Criterion

Remark 2: Let $\mathscr{G}^{\sigma} \triangleq \{G(s) \mid \operatorname{Re} s = \sigma\}$. It can be shown that it suffices for instability that $d(\mathscr{D}, \mathscr{G}^{\sigma}) > 0$ for some $\sigma \geqslant 0$ and that $d(\mathscr{D}, \mathscr{G}^{+}) = 0$. This in fact leads to an improved instability criterion[5] in situations such as the one illustrated in Figure 5.6.

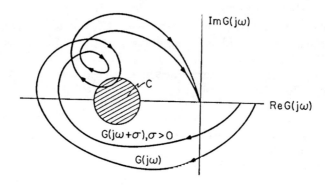

Figure 5.6   Illustration of Remark 2

References

1. Desoer, C. A., "A General Formulation of the Nyquist Criterion," *IEEE Trans. on Circuit Theory*, Vol. CT-12, pp. 230–234, 1965.
2. Desoer, C. A., and Wu, M. Y., "Stability of Linear Time-Invariant Systems," *IEEE Trans. on Circuit Theory*, Vol. CT-15, pp. 245–250, 1968.
3. Desoer, C. A., and Wu, M. Y., "Stability of Multiple-Loop Feedback Linear Time-Invariant Systems," *J. of Math. Analysis and Applications*, Vol. 23, No. 1, pp. 121–129, 1968.
4. Desoer, C. A., and Wu, M. Y., *Input-Output Properties of Multiple-Input Multiple-Output Discrete Systems, Part I*, Memorandum No. ERL-M259, University of California, Berkeley, 1969.
5. Hille, E., and Phillips, R. S., *Functional Analysis and Semi-Groups* (second edition), American Mathematical Society, Providence, 1957.
6. Zames, G., "On the Input-Output Stability of Time-Varying Nonlinear Feedback Systems. Part I: Conditions Derived Using Concepts of Loop Gain, Conicity, and Positivity. Part II: Conditions Involving Circles in

---

[5] It has been conjectured that the instability part of the circle criterion could be extended to yield instability whenever every point of the critical disc belongs to the spectrum of the forward-loop operator. In terms of encirclements this would imply instability if every point of the critical disk is encircled at least once. (The present instability theorem requires all points to be encircled the same number of times.) This conjecture is due to Professor R. W. Brockett of Harvard University.

the Frequency Plane and Sector Nonlinearities," *IEEE Trans. on Automatic Control*, Vol. AC-11, pp. 228–238 and 465–476, 1966.
7. Rudin, W., *Real and Complex Analysis*, McGraw-Hill, New York, 1966.
8. Sandberg, I. W., "On the L_2-Boundedness of Solutions of Nonlinear Functional Equations," *Bell System Tech. J.*, Vol. 43, pp. 1581–1599, 1964.
9. Sandberg, I. W., "Some Results on the Theory of Physical Systems Governed by Nonlinear Functional Equations," *Bell System Tech. J.*, Vol. 44, pp. 871–898, 1965.
10. Zames, G., "On the Stability of Nonlinear, Time-Varying Feedback Systems," *Proc. 1964 Natl. El. Conf.*, Vol. 20, pp. 725–730, 1964.
11. Narendra, K. S., and Goldwyn, R. M., "A Geometrical Criterion for the Stability of Certain Nonlinear Nonautonomous Systems," *IEEE Trans. on Circuit Theory*, Vol. CT-11, pp. 406–408, 1964.
12. Kudrewicz, "Stability of Nonlinear Systems with Feedback," *Automation and Remote Control*, Vol. 25, pp. 1027–1037, 1964.
13. Brockett, R. W., and Lee, H. B., "Frequency-Domain Instability Criteria for Time-Varying and Nonlinear Systems," *IEEE Proceedings*, Vol. 55, pp. 604–619, May 1967.
14. Willems, J. C., "Stability, Instability, Invertibility and Causality," *SIAM J. on Control*, Vol. 7, No. 4, pp. 645–671, 1969.

6 Stability Criteria Obtained Using Multipliers

6.1 Introduction

There are two basic transformations of feedback systems which lie at the foundation of most of the frequency-domain stability criteria as they have recently appeared in the control theory literature.

The first one is the loop transformation shown in Figure 6.1. This transformation results in a shift of the conicity of the operators in the forward and the feedback loop and can, for instance, be used to transform a feedback system in which the forward and the feedback loop satisfy certain conicity conditions into a feedback system which is open-loop attenuating. Corollary 4.3.3 in fact rests on this principle. This procedure can also result in a feedback system with positive operators in both the forward and the feedback loop.

The second basic transformation is the introduction of so-called multipliers in the loop.[1] This is illustrated in Figure 6.2 and results in the possibility of exploiting certain constraints of the operators in the loop. Assume, for instance, that the operator in the feedback loop is conic and satisfies some additional conditions — for example, that it is time invariant and memoryless (but nonlinear) or that it is a linear periodic gain. In general, it is then possible to find a multiplier which

---

[1] The idea of using multipliers can be traced back to the work of Popov (Ref. 1). Other authors who developed this technique are Brockett and Willems (Ref. 2) and Zames (Ref. 3).

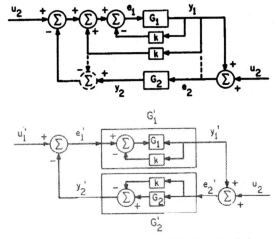

Figure 6.1 Transformations of the Feedback Loop

when cascaded with this operator will not change the conicity of the feedback loop, whereas the compensation of this multiplier in the forward loop will change the conicity of the forward loop, thus making it possible to show stability whereas this was *not* possible before the conicities were changed by means of the multipliers.

The above two transformations are usually used in series in the sense that the conicity transformation is used first and results in making the operator in the feedback loop (which is assumed to satisfy certain constraints) a positive operator.[2] The forward loop is in this process appropriately modified so as to preserve the original input-output relationship. One then introduces a multiplier in cascade with the operator in the feedback loop. This multiplier is appropriately chosen

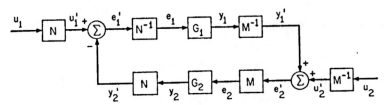

Figure 6.2 Illustration of the Introduction of Multipliers

[2] There is no a priori reason to make these transformations in precisely this sequence, although some theoretical considerations supporting this sequence can be made (Ref. 4).

out of a certain class without affecting the positivity of the feedback loop. If the compensation of this multiplier results in a positive operator in the forward loop, then stability results.

These ideas are further explored in the remainder of this chapter. They will be investigated first from a theoretical point of view and then applied to two representative cases. The first case treats feedback systems with a periodic gain in the feedback loop, and the second case treats feedback systems with a monotone nonlinearity in the feedback loop.

6.2 Transformations of the Feedback Loop

The feedback loop referred to in this section is described by the functional equations (FE) as introduced in Section 4.2. These are restated here for convenience:

$$e_1 = u_1 - y_2,$$

$$e_2 = u_2 + y_1,$$

$$y_1 = G_1 e_1,$$

$$y_2 = G_2 e_2.$$

(FE)

It is still assumed that the assumptions W.1 through W.4, G.1 through G.4, and I.1 are satisfied. These assumptions are enumerated in Section 4.2.

THEOREM 6.1

Consider the feedback system described by equations (FE) and assume that $W_1 = W_2$. Let $k \neq 0$ be a scalar such that $-k^{-1} \notin \sigma(G_1)$ in $\mathcal{N}^+(W_{1e}, W_{1e})$ (i.e., $I + kG_1$ has a causal inverse on W_{1e}). Let $G_1' = G_1(I + kG_1)^{-1}$ and $G_2' = G_2 - kI$. Let $u_1 \in W_{1e}$ and $u_2 \in W_{2e}$ be given and assume that $\{e_1, e_2, y_1, y_2\}$ is a corresponding solution. Then the quadruple

$$\{e_1' = e_1 + ky_1, \, e_2' = e_2, \, y_1' = y_1, \, y_2' = y_2 - ke_2\}$$

satisfies the equations

$$e_1' = u_1' - y_2',$$

$$e_2' = u_2' + y_1',$$

$$y_1' = G_1' e_1',$$

$$y_2' = G_2' e_2',$$

(FE')

where $u_1' = u_1 - ku_2$ and $u_2' = u_2$. Moreover, if the feedback systems described by the equations (FE) and (FE') are both well posed, then (finite gain) stability (continuity, Lipschitz continuity) of one implies and is implied by (finite gain) stability (continuity, Lipschitz continuity) of the other.

Proof: The first part of the theorem is readily proved by direct verification, whereas the stability claim is a direct consequence of the relation between the solutions, since this relation is easily reversed from (FE') to (FE).

THEOREM 6.2

Let $N \in \mathscr{B}^+(W_{1e}, W_{1e})$ ($\tilde{\mathscr{B}}^+(W_{1e}, W_{1e})$) be regular in $\mathscr{B}^+(W_{1e}, W_{1e})$ ($\tilde{\mathscr{B}}^+(W_{1e}, W_{1e})$) and $M \in \mathscr{B}^+(W_{2e}, W_{2e})$ ($\tilde{\mathscr{B}}^+(W_{2e}, W_{2e})$) be regular in $\mathscr{B}^+(W_{2e}, W_{2e})$ ($\tilde{\mathscr{B}}^+(W_{2e}, W_{2e})$), i.e., M, M^{-1}, N, N^{-1} are causal bounded (Lipschitz-continuous) operators. Consider now the feedback system described by the equations

$$e_1'' = u_1'' - y_2'',$$
$$e_2'' = u_2'' + y_1'',$$
$$y_1'' = M^{-1}G_1N^{-1}e_1', \tag{FE''}$$
$$y_2'' = NG_2Me_2'.$$

Assume that both feedback systems (FE) and (FE'') are well posed. Then the feedback system described by the equations (FE) is (finite gain) stable (continuous, Lipschitz continuous) if and only if the feedback system described by the equations (FE'') is (finite gain) stable (continuous, Lipschitz continuous).

Proof: It is a simple matter to verify that the solutions to the feedback systems are related through $u_1'' = Nu_1$, $e_1'' = Ne_1$, $y_1'' = M^{-1}y_1$, $u_2'' = M^{-1}u_2$, $e_2'' = M^{-1}e_2$, $y_2'' = Ny_2$, which then in view of the assumptions on M and N readily yields the theorem.

Theorem 6.2 immediately leads to the following useful corollary.

COROLLARY 6.2.1

Let $W_1 = W_2$ be Hilbert spaces. Let $M, N \in \mathscr{B}^+(W_{1e}, W_{1e})$ ($\tilde{\mathscr{B}}^+(W_{1e}, W_{1e})$) be regular in $\mathscr{B}^+(W_{1e}, W_{1e})$ ($\tilde{\mathscr{B}}^+(W_{1e}, W_{1e})$). Then the feedback system described by equations (FE) is finite gain stable

(Lipschitz continuous) if it is well posed, if $G_2' = NG_2M$ is (incrementally) positive on W_{2e} and if $G_1' = M^{-1}G_1N^{-1}$ is strictly (incrementally) positive and Lipschitz continuous on W_1.

Proof: This corollary is an immediate consequence of Theorem 6.2 and Corollary 4.3.4.

As shown in Chapter 3, one often obtains positive operators using noncausal multipliers. Since such multipliers would violate the causality condition of Corollary 6.2.1, it becomes necessary to introduce factorizations. This is the subject of the following corollary.

COROLLARY 6.2.2

Let $W_1 = W_2$ be Hilbert spaces. Assume that $Z \in \tilde{\mathscr{B}}(W_2,W_2)$ admits a factorization into $Z = MN$ with

1. $M \in \mathscr{L}(W_2,W_2)$ is invertible, and M^*, $(M^{-1})^* \in \mathscr{L}^+(W_2,W_2)$ (i.e., M and M^{-1} are anticausal bounded linear operators);
2. $N \in \mathscr{B}^+(W_2,W_2)$ is invertible, and $N^{-1} \in \mathscr{B}^+(W_1,W_1)$; and
3. ZG_2 is (incrementally) positive on W_2 and G_1Z^{-1} is strictly (incrementally) positive and Lipschitz continuous on W_1.

Then the feedback system described by equations (FE) is finite gain stable (Lipschitz continuous) if it is, in addition, well posed.

Proof: Since MNG_2 is positive on W_2, $NG_2(M^{-1})^*$ is positive on W_2 and since $G_1N^{-1}M^{-1}$ is strictly positive and Lipschitz continuous on W_1, $M^*G_1N^{-1}$ is strictly positive and Lipschitz continuous on W_1. The theorem then follows from Corollary 6.2.1.[3]

6.3 A Stability Criterion for Linear Feedback Systems with a Periodic Gain in the Feedback Loop

The ideas presented in Section 6.2 and the results obtained in Chapters 4 and 5 will now be applied to a particular class of feedback systems. This class consists of feedback systems with a linear time

[3] The theorems and corollaries of this section can be sharpened if one assumes that the preliminary conditions of Theorem 4.3 to be satisfied. Some such results are stated in Reference 5. The most significant relaxation is that it is then possible to state theorems involving a factorization of G_2G_1 rather than requiring this factorization be of the type $G_2MM^{-1}G_1$.

invariant operator in the forward loop and a periodic gain in the feedback loop.[4] The resulting system is shown in Figure 6.3.

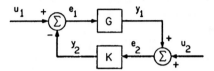

Figure 6.3   The Feedback System under Consideration in Section 6.3

In terms of the notation used in the previous chapters, let the time-interval of definition $S = [T_0, \infty)$, $V_1 = V_2 = R$, i.e., u_1, e_1, y_1, u_2, e_2, y_2 are real-valued functions of time on $[T_0, \infty)$, and let the operators G and K be formally defined by

$$(Gx)(t) \triangleq \sum_{n=0}^{\infty} g_n x(t - t_n) + \int_{T_0}^{\infty} g(t - \tau)x(\tau)\, d\tau,$$

$$(Kx)(t) \triangleq k(t)x(t),$$

where $\{t_n\}$, $n \in I^+$, is a sequence of real numbers with $t_0 = 0$, $t_n > 0$ for $n \geqslant 1$, $g(t)$ is a real-valued function on R with $g(t) = 0$ for $t < 0$ and $g \in L_1(0, \infty)$, $\{g_n\} \in l_1$ is a real-valued sequence, and $k(t)$ is a real-valued function on $[T_0, \infty)$ with $k(t + T) = k(t)$, for some $T > 0$ and all $t \geqslant T_0$, with $k(t) \in L_\infty(T_0, \infty)$. The functional equations describing the feedback system are:

$$e_1(t) = u_1(t) - y_2(t),$$

$$e_2(t) = u_2(t) + y_1(t),$$

$$y_1(t) = (G_1 e_2)(t), \text{ and} \qquad \text{(LPFE)}$$

$$y_2(t) = (G_2 e_2)(t).$$

The solution space will be taken as $L_{2e}(T_0, \infty)$. The inputs u_1, $u_2 \in L_{2e}(T_0, \infty)$. The above operators have been studied in detail in Chapter 3. It follows readily from Minkowski's inequality and some direct estimates that G and K map $L_{2e}(T_0, \infty)$ into itself. They are in fact both Lipschitz-continuous linear operators on $L_{2e}(T_0, \infty)$. The operator G

---

[4] There exists a large mathematical literature on the stability of linear differential equations with periodic coefficients. The specific results obtained here evolved out of the so-called "pole-following technique" ingeniously introduced by Bongiorno (Refs. 6, 7) and exploited, among others, by Sandberg (Ref. 8) and the author (Ref. 9.)

is time invariant. The feedback system described by the equation (LPFE) thus satisfies the assumptions W.1 through W.4, G.1 through G.4, and I.1, and the theory developed in Chapter 4 is applicable. Notice that the conditions on g, $\{g_n\}$, and $k(t)$ imply open-loop stability of the feedback system.

As an example of this, consider the linear time-varying ordinary differential equation

$$p(D)x(t) + k'(t)q(D)x(t) = 0, \qquad D^i = \frac{d^i}{dt^i}.$$

The following assumptions are made:

A.1. The functions $p(s)$ and $q(s)$ are real polynomials in s, i.e.,

$$p(s) = s^n + p_{n-1}s^{n-1} + \cdots + p_0,$$

$$q(s) = q_n s^n + q_{n-1}s^{n-1} + \cdots + q_0,$$

with p_i and q_i real numbers.

A.2. The function $k'(t)$ is a real-valued piecewise continuous function of t and belongs to L_∞.

A.3. Either of the following conditions is satisfied:

1. $q_n = 0$; or
2. $q_n \neq 0$ and $-1/q_n \notin [\alpha,\beta]$,
   where α and β are such that $\alpha \leqslant k'(t) \leqslant \beta$ for all $t \in R$.

A real-valued continuous function $x(t)$ is said to be a solution of this time-varying differential equation if it possesses $(n-1)$ continuous derivatives and if it satisfies the differential equation for all t for which $k'(t)$ is continuous. Clearly $x(t) \equiv 0$ is a solution. This solution is called the *null solution* and is said to be *asymptotically stable* if all solutions approach the null solution for $t \to \infty$.

Asymptotic stability of the null solution of the time-varying differential equation can be deduced from L_2-stability of a feedback system of the type which is being considered in this section. Assume therefore that there exists a real number α such that the zeros of the polynomial $p(s) + \alpha q(s)$ have a negative real part.

It can be shown without much difficulty (see Ref. 9) that the differential equation can be rewritten as

$$p_1(D)x(t) + k_1(t)q_1(D)x(t) = 0,$$

with $p_1(s)$ a monic Hurwitz polynomial of degree n (i.e., all its zeros have a negative real part, and the coefficient of s^n is one) with the degree of $p_1(s)$ larger than the degree of $q_1(s)$. This nth order scalar differential equation is equivalent to the first-order vector differential equation

$$\frac{dz(t)}{dt} = Az(t) + bu(t),$$

with

$$y(t) = c'z(t)$$

and

$$u(t) = k_1(t)y(t),$$

where

$$z(t) = \operatorname{col}\left(x(t), \frac{dx(t)}{dt}, \ldots, \frac{d^{n-1}x(t)}{dt^{n-1}}\right),$$

$$A = \begin{bmatrix} 0 & 1 & 0 & \cdots & 0 \\ 0 & 0 & 1 & \cdots & 0 \\ \cdot & \cdot & \cdot & & \\ \cdot & \cdot & \cdot & & \\ \cdot & \cdot & \cdot & & \\ 0 & 0 & 0 & \cdots & 1 \\ -p_{1,0} & -p_{1,1} & -p_{1,2} & \cdots & -p_{1,n-1} \end{bmatrix},$$

$$b = \operatorname{col}(0, 0, \ldots, 0, 1),$$

$$c = \operatorname{col}(q_{1,0}, q_{1,1}, \ldots, q_{1,n-1}),$$

$$c'(Is - A)^{-1}b = q_1(s)/p_1(s).$$

The null solution of the differential equation under consideration will then be asymptotically stable if and only if given any $z(0)$, $\lim_{t\to\infty} \|z(t)\|$ exists and is zero. It is well known that the smoothness conditions on $k'(t)$ are sufficient to ensure the existence of a unique solution which assumes the value $z(0)$ for $t = 0$. Furthermore, the solutions satisfy the integral equation

$$z(t) = e^{At}z(0) - \int_0^t e^{A(t-\tau)}bk_1(\tau)y(\tau)\,d\tau \qquad \text{for} \qquad t \geqslant 0,$$

which implies that

$$y(t) = c'e^{At}z(0) - \int_0^t c'e^{A(t-\tau)}bk_1(\tau)y(\tau)\,d\tau \qquad \text{for} \qquad t \geqslant 0.$$

It is clear that this last equation represents the feedback system of Figure 6.3 with $T_0 = 0$,

$$u_1(t) = 0, \qquad\qquad\qquad u_2(t) = r(t) \triangleq c'e^{At}z(0)$$

$$\text{for} \quad t \geqslant 0,$$

$$y_1(t) = y(t), \qquad\qquad y_2(t) = k_1(t)(y(t) + r(t)),$$

$$e_1(t) = -y_2(t), \qquad\qquad e_2(t) = y(t) + r(t),$$

$$g(t) = \begin{cases} c'e^{At}b & \text{for} \quad t \geqslant 0, \\ 0 & \text{otherwise,} \end{cases} \qquad k(t) = k_1(t), \quad \text{and}$$

$$g_k = 0 \quad \text{for all} \quad k \in I.$$

It follows from the assumption on the zeros of $p_1(s)$ that all eigenvalues of A have a negative real part and thus that $c'e^{At}b \in L_p(0,\infty)$ for all $p \geqslant 1$.

Thus, proof of L_2-stability for the feedback system implies that all solutions $z(t)$ to the vector differential equation which are such that $y \in L_{2e}(0,\infty)$ also belong to $L_2(0,\infty)$. Since all solutions $z(t)$ are continuous, all solutions $y(t)$ do belong to $L_{2e}(0,\infty)$ and hence all solutions yield $y \in L_2(0,\infty)$. Since

$$z(t) = e^{At}z(0) + \int_0^t e^{A(t-\tau)}bk_1(\tau)y(\tau) \, d\tau$$

and $e^{At}b \in L_1^{R^n}(0,\infty)$, $k_1 \in L_\infty(0,\infty)$, $e^{At}z(0) \in L_2(0,\infty)$ and the convolution of an L_1-function with an L_2-function yields an L_2-function, it follows thus that $z \in L_2(0,\infty)$. Furthermore

$$\frac{dz(t)}{dt} = Az(t) - k_1(t)bc'z(t),$$

hence $dz/dt \in L_2(0,\infty)$. Since z and dz/dt belong to $L_2(0,\infty)$, $\lim_{t\to\infty} z(t)$ exists and is zero. Hence, L_2-stability of the above feedback system implies asymptotic stability of the null solution of the differential equation.

These simple manipulations show that although it might at first glance seem that the type of stability which is obtained in the theorem in the previous section is not as strong as Lyapunov stability, in most circumstances it actually implies it.

Notice that the periodicity of $k(t)$ was not used in the preceding argument. Using the periodicity of $k(t)$ it can in fact be shown by invoking Floquet theory (Ref. 10) that if the system described by equation (LPFE) is L_2-stable, then it is L_p-stable for any $1 \leqslant p \leqslant \infty$.

Feedback systems of the type described by the functional equations (LPFE) or by the ordinary differential equation occur frequently in the design of systems containing parametric devices. The stability properties of such systems are of course of primary importance, and criteria using frequency-domain conditions similar to the Nyquist criterion have proven to be a particularly useful tool. Moreover, the local stability of a periodic solution of a nonlinear differential equation is equivalent to the stability of the null solution of a linear time-varying differential equation of the form illustrated here.

The stability properties of the feedback system under consideration have received a great deal of attention in the past, and the result that is best known is the circle criterion, which was discussed in Chapter 5. Although the circle criterion is applicable under much weaker conditions (the feedback gain need *not* be linear or periodic) than the ones stated on p. 141, it was originally proved making essentially the same assumptions.

In this section a new frequency-domain stability criterion is developed which assumes explicitly that the feedback gain is linear and periodic with a certain *given* period. This assumption makes it then possible to obtain an improved stability criterion. The result gives — for a particular transfer function of the forward loop — combinations of the lower bound α, the upper bound β, and the period T of $k(t)$ that yield stability. This dependence on the period is of course as expected and has been investigated exhaustively for certain classical types of second-order differential equations. The criterion to be stated in Theorem 6.3 requires, as do most recent frequency-domain stability criteria, the existence of a multiplier having certain properties. With the exception of the Popov criterion, however, there is generally no procedure offered to determine whether or not such a multiplier exists for a given transfer function of the forward loop (see Ref. 3). This is not the case for the criterion presented here, since Theorem 6.3 can be completely rephrased in terms of this transfer function. In fact, a simple graphical procedure is given to determine whether or not the multiplier exists.

THEOREM 6.3

Assume that the feedback system described by the functional equations (LPFE) is well posed, and that:

1. $\alpha + \epsilon \leqslant k(t) = k(t + T) \leqslant \beta - \epsilon$ for some $\epsilon > 0$ and almost all $t \geqslant T_0$;

2. $\inf_{\text{Re } s \geqslant 0} |1 + KG(s)| > 0$ for some $K \in [\alpha, \beta]$, where $G(s)$ denotes the Laplace transform of $(g, \{g_n, t_n\})$; and

3. there exists a complex-valued function $F(j\omega)$ of the real variable, ω, with $\overline{F(j\omega)} = F(-j\omega)$ such that for almost all $\omega \geqslant 0$

   F.1. $\text{Re } \{F(j\omega)\} \geqslant \epsilon$ for some $\epsilon > 0$,

   F.2. $F(j\omega) = F(j(\omega + 2\pi T^{-1})) \in L_\infty$,   and

   F.3. $\text{Re } \left\{ F(j\omega) \dfrac{\beta G(j\omega) + 1}{\alpha G(j\omega) + 1} \right\} \geqslant 0.$

Then the feedback system is L_2-stable.

Proof: It can be shown that the condition of the theorem implies that $\inf_{\text{Re } s \geqslant 0} |1 + KG(s)| > 0$ for any $K \in [\alpha, \beta]$. The remainder of the proof is divided into three steps.

A. Consider the transformation of the feedback loop shown in Figure 6.1 and treated in Theorem 6.1 with $k = \alpha$ and the roles of the forward and the feedback loop reversed (i.e., the feedforward is taken around the feedback path). It thus suffices to prove stability for the system with $G' = G(I + \alpha G)^{-1}$ in the forward loop and the gain $k(t) - \alpha$ in the feedback loop. This transformation is now repeated with the roles of the operators reversed (i.e., the feedforward is now around the forward path) and $k = -1/(\beta - \alpha)$. It thus suffices to prove stability for the system with $G'' = G(I + \alpha G)^{-1} + [1/(\beta - \alpha)]I$ in the forward loop and the gain $(k(t) - \alpha)(1 - [1/(\beta - \alpha)](k(t) - \alpha))^{-1}$ in the feedback. A simple manipulation now shows that

$$G'' = \frac{1}{\beta - \alpha} \frac{I + \beta G}{I + \alpha G}$$

and that the gain in the feedback loop equals

$$k''(t) = (\beta - \alpha) \frac{k(t) - \alpha}{\beta - k(t)}.$$

B. The assumptions on $F(j\omega)$ do not suffice to ensure that it has a Fourier series in l_1. However, since $G(j\omega)$ is uniformly continuous for $-\infty < \omega < \infty$ and $\lim_{|\omega| \to \infty} G(j\omega) = 0$, it follows that $F(j\omega)$ may always be taken to have a Fourier series in l_1, say $\{f_n\}$, $n \in I$, (in fact, it may be assumed that $F(j\omega)$ has a finite Fourier series).

C. Since $\text{Re } F(j\omega) \geqslant \epsilon > 0$, $F^{-1}(j\omega) \in L_\infty$. It is now a simple matter to verify that the operator defined on $L_2(-\infty, +\infty)$ by $(Zx)(t) = \sum_{n \in I} f_n x(t - nT)$ is invertible and that by Theorem 3.6 the operators $G''Z^{-1}$ and ZK'' are respectively positive and strictly positive bounded linear operators on $L_2(-\infty, +\infty)$. The manipulations as in Corollary 3.2.2 and the factorization of Theorem 3.20 now show that the conditions of Corollary 6.2.2 are satisfied, which thus yields stability as claimed. The fact that the operators can be considered as operators on $L_2(T_0, \infty)$ follows from the causality of the operators after the proper factorizations have been carried out.

Theorem 6.3 is not very useful as it stands since it leaves un-answered the question whether or not the multiplier $F(j\omega)$ exists. This question can be resolved, however, and this leads to an equivalent formulation of the theorem.
Let

$$\phi_{\max}(\omega) = \sup_{n \in I} \phi(\omega + n2\pi T^{-1}),$$

$$\phi_{\min}(\omega) = \inf_{n \in I} \phi(\omega + n2\pi T^{-1})$$

where

$$\phi(\omega) = \arg \frac{\beta G(j\omega) + 1}{\alpha G(j\omega) + 1}.$$

The alternate formulation of Theorem 6.3 is then:

THEOREM 6.3A
Assume that the feedback system described by the functional equations (LPFE) is well-posed, and that:

1. $\alpha + \epsilon \leqslant k(t) = k(t + T) \leqslant \beta - \epsilon$ for some $\epsilon > 0$ and almost all $t \geqslant T_0$;
2. $\inf_{\mathrm{Re}\ s \geqslant 0} |1 + KG(s)| > 0$ for some $K \in [\alpha, \beta]$, where $G(s)$ denotes the Laplace transform of $(g, \{g_n, t_n\})$; and
3. $\phi_{\max}(\omega) - \phi_{\min}(\omega) < \pi$ for all $|\omega| \leqslant \pi T^{-1}$.
Then the feedback system is L_2-stable.

Proof: Since $G(j\omega)$ is a uniformly continuous and bounded function of ω, the sequence of functions $G(j\omega + n2\pi T^{-1})$, $n \in I$, is equicontinuous and thus $\phi_{\max}(\omega)$ and $\phi_{\min}(\omega)$ are continuous functions of ω. Hence, $|\phi_{\max}(\omega) - \phi_{\min}(\omega)|$ is a continuous function of ω. Since by symmetry $\phi_{\max}$ and $\phi_{\min}$ are periodic,

$$\phi_{\max}(\omega) - \phi_{\min}(\omega) = \phi_{\max}(\omega + 2\pi T^{-1}) - \phi_{\min}(\omega + 2\pi T^{-1}).$$

Since $|\phi_{\max}(\omega) - \phi_{\min}(\omega)| < \pi$, there exists an $\epsilon > 0$ such that $|\phi_{\max}(\omega) - \phi_{\min}(\omega)| \leqslant \pi - \epsilon$. Let

$$F(j\omega) = \exp\{-\tfrac{1}{2}j[\phi_{\max}(\omega) + \phi_{\min}(\omega)]\}.$$

It is easily verified that this choice for $F(j\omega)$ yields the conclusion by Theorem 6.3. For the converse part of the equivalence, assume that $\phi_{\max}(\omega') - \phi_{\min}(\omega') = \pi$ for some $\omega' \in R$. Then, since Re $\{G(j\omega)F(j\omega)\}$ has to be nonnegative for all ω, this implies that $|\arg F(j\omega')| \geqslant \pi/2$, which contradicts the condition that Re $F(j\omega) \geqslant \epsilon > 0$.

The following two corollaries show that the criterion is a trade-off between the circle criterion (T arbitrary) and the "frozen time" Nyquist criterion (T small).

COROLLARY 6.3.1

The feedback described by the equations (LPFE) is L_2-stable if it is well posed, if $k(t)$ is periodic, and if:

1. $\alpha + \epsilon \leqslant k(t) \leqslant \beta - \epsilon$ for some $\epsilon > 0$ and almost all $t \geqslant T_0$;
2. $\inf_{\mathrm{Re}\ s \geqslant 0} |1 + KG(s)| > 0$ for some $K \in [\alpha,\beta]$ where $G(s)$ denotes the Laplace transform of $(g,\{g_n,t_n\})$; and
3. Re $\dfrac{\beta G(j\omega) + 1}{\alpha G(j\omega) + 1} \geqslant 0.$

Proof: Take $F(j\omega) = 1$ and apply Theorem 6.3.

Corollary 6.3.1 is a particular case (since it assumes the feedback gain to be linear and periodic) of the circle criterion.

Consider now the stability properties of the linear time-invariant system obtained by replacing $k(t)$ in the feedback loop by $k_t = k(t)$ for some t. Even though the time-invariant system thus obtained is L_2-stable for all constants k_t, it does not follow in general that the original feedback system is L_2-stable (see Ref. 11). This fact is closely related to the Aizerman conjecture for time-invariant systems to be discussed in the next chapter. However, the following corollary shows that this procedure is legitimate if the period T is sufficiently small. The corollary states that if the frequency of the feedback gain is sufficiently high compared to the natural frequencies of the forward loop then no instability due to "pumping" can occur.[5]

---

[5] This result is not the sharpest one possible. Indeed, it can be shown (Ref. 10) that for the period sufficiently small, it suffices that the constant feedback system be stable when $k(t)$ is replaced by its average value.

COROLLARY 6.3.2

Assume, in the definition of G, that $g_n = 0$ for all $n \geqslant 1$ and that the feedback system is L_2-stable for any $k(t) = k = $ constant in the feedback loop with $\alpha \leqslant k \leqslant \beta$. Then there exists a T_1 such that for all $T < T_1$ the feedback system with any gain $\alpha \leqslant k(t) = k(t + T) \leqslant \beta$ in the feedback loop is also L_2-stable.

Proof: Since $\lim_{|\omega| \to \infty} G(j\omega) = g_0$ exists (by the Riemann-Lebesgue lemma) and is real, $\lim_{|\omega| \to \infty} \phi(\omega)$ exists and is zero. Since the feedback system is L_2-stable for constant gains k in the feedback loop with $\alpha \leqslant k \leqslant \beta$, there exists a function of $Z(j\omega)$ such that for all ω,

$$\text{Re } Z(j\omega) \geqslant \epsilon > 0 \quad \text{and} \quad \text{Re } Z(j\omega) \frac{\beta G(j\omega) + 1}{\alpha G(j\omega) + 1} \geqslant 0.$$

(This follows from the Nyquist diagram and a simple graphical construction.) It thus follows that for ω_0 sufficiently large the function $F(j\omega) = Z(j\omega)$ for $|\omega| \leqslant \omega_0/2$ and $F(j\omega) = F(j(\omega + \omega_0))$ otherwise, will yield the conclusion by Theorems 6.3.

9.3.1 Application of the Criterion

Theorem 6.3A suggests an obvious graphical procedure for determining whether or not Theorem 6.3 predicts L_2-stability. This is illustrated in Figure 6.4, and requires plotting the curves $\phi_N(\Omega) = \phi(\Omega + N\omega_0)$, $\omega_0 = 2\pi T^{-1}$, versus Ω for $|\Omega| \leqslant \omega_0/2$ and $N \in I$. The

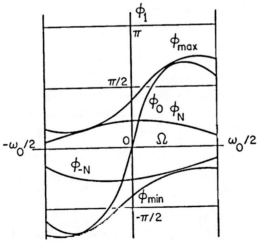

Figure 6.4 Graphical Procedure for Determining $F(j\omega)$

upper and lower envelope of these curves give $\phi_{max}(\Omega)$ and $\phi_{min}(\Omega)$. Theorem 6.3 then requires for L_2-stability that $\phi_{max}(\Omega) - \phi_{min}(\Omega) < \pi$ for all $|\Omega| \leqslant \omega_0/2$. It is apparent that this procedure, although straightforward, is rather tedious.

In order to facilitate the application of the criterion, some simple *necessary* conditions for the multiplier $F(s)$ to exist are now given for the case $0 < \alpha \leqslant \beta$:

1. The Nyquist locus of $G(s)$ should not encircle or intersect the straight line segment $[-1/\alpha, -1/\beta]$ of the negative real axis of the Nyquist plane.
2. The points $G(jn\omega_0/2)$, $n = 0, 1, 2, \ldots$, should satisfy the conditions of the circle criterion; i.e., they should not lie inside the closed disk centered on the negative real axis at $-\frac{1}{2}(1/\alpha + 1/\beta)$ with radius $\frac{1}{2}(1/\alpha - 1/\beta)$.

Analogous conditions hold for other ranges of α and β.

The second necessary condition follows from the fact that, since $F(j\omega) = F(-j\omega)$, and since $F(j(\omega + \omega_0)) = F(j\omega)$, then

$$F(jn\omega_0/2) = \text{Re } F(jn\omega_0/2)$$

for $n = 0, \pm1, \pm2, \ldots$. Thus conditions F.1 and F.3 of Theorem 6.3 imply that

$$\text{Re } \frac{\beta G(jn\omega_0/2) + 1}{\alpha G(jn\omega_0/2) + 1} > 0 \quad \text{for} \quad n \in I,$$

which leads to the second necessary condition.

By choosing particular functions for $F(j\omega)$ it is of course possible to obtain other sufficient conditions for L_2-stability. The next corollary is based on this idea and gives a quite simple *sufficient* condition for the multiplier $F(j\omega)$ to exist. It is expressed entirely in terms of the Nyquist locus of $G(s)$, and is stated here for the case $0 < \alpha \leqslant \beta$.

COROLLARY 6.3.3

Assume, in the definitions of G, that $t_n = nT_1$ for some $T_1 > 0$. Then the feedback system described by equations (LPFE) is L_2-stable if it is well posed, and if:

1. the Nyquist locus of $G(s)$ does not encircle the point $-1/\alpha$ on the negative real axis of the Nyquist plane; and

2. there exists a circle, C, which passes through the points $-1/\alpha$ and $-1/\beta$, such that the Nyquist locus of $G(s)$ for $\omega \geqslant 0$ does not intersect it.

Let C' be the mirror image of C with respect to the real axis, and consider the following two parts of the Nyquist locus of $G(s)$:

$$S_1: \{G(j\omega) \mid n\omega_0 \leqslant \omega \leqslant (n + 1/2)\omega_0\}, \text{ and}$$

$$S_2: \{G(j\omega) \mid (n + 1/2)\omega_0 \leqslant \omega \leqslant (n + 1)\omega_0\},$$

where $n \in I^+$. Then

3. C' does not intersect both S_1 and S_2.

This corollary is illustrated in Figure 6.5.

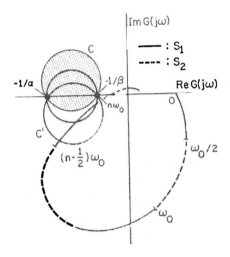

Figure 6.5   Illustration of Corollary 3.3

Proof: Condition 1 assures that the third condition of Theorem 6.3 is satisfied. Let $|\theta| \leqslant \pi/2$ be the angle between the positive real axis and the straight line through the origin of the complex plane defined by the points

$$\left\{ \left| \frac{\beta\tau + 1}{\alpha\tau + 1} \right| \tau \in C \right\}.$$

Assume that $\theta \geqslant 0$ and that C' does not intersect S_2 (a similar argument establishes the corollary for the other cases). Let $F(j\omega)$ be a function of ω such that for $n \in I$

$$\arg F(j\omega) = \begin{cases} \pi/2 - \theta & \text{for} \quad n\omega_0 < \omega < (n + 1/2)\omega_0 \\ -(\pi/2 - \theta) & \text{for} \quad (n - 1/2)\omega_0 < \omega < n\omega_0. \\ 0 & \text{for} \quad \omega = n\omega_0,\ (n + 1/2)\omega_0 \end{cases}$$

Clearly, $F(j\omega)$ satisfies conditions F.1 and F.2 of Theorem 6.3. From condition 1 of the corollary it follows that

$$-\pi + \theta < \phi(\omega) < \theta$$

for $\omega \geqslant 0$ and $n\omega_0 < \omega < (n + 1/2)\omega_0$; and from the fact that C' does not intersect S_2 it follows that

$$-\theta < \phi(\omega) < \pi - \theta$$

for $\omega \geqslant 0$ and $(n - 1/2)\omega_0 \leqslant \omega \leqslant n\omega_0$. Thus it follows that

$$-\pi/2 < \arg F(j\omega) + \phi(\omega) < \pi/2$$

for $\omega \geqslant 0$, which establishes condition F.3 of Theorems 6.3 since $\arg F(-j\omega) + \phi(-\omega) = -\arg F(j\omega) - \phi(\omega)$.

The following two examples illustrate the usefulness of Theorem 6.3 and its corollaries.

1. Let

$$G(s) = \frac{s}{(s + 10)(s^2 + 0.4s + 1)}.$$

$k(t) = k(t + T)$ and $0 \leqslant k(t) \leqslant 2$. Determine for which range of

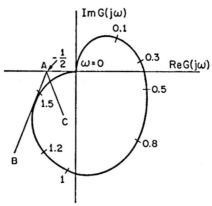

Figure 6.6   Nyquist Locus of $s/(s + 10)(s^2 + 0.4s + 1)$

$\omega_0 = 2\pi/T$ this feedback system is stable. The Nyquist locus of $G(s)$ is shown in Figure 6.6.

It is apparent from the Nyquist locus that the circle criterion cannot be used to predict L_2-stability. Using the procedure suggested at the beginning of this section, Theorem 6.3 shows that this feedback system is L_2-stable for all $k(t)$ in the determined range provided $\omega_0 > 1.55$. Using Corollary 6.3.3, on the other hand, this feedback system is found to be L_2-stable for all $k(t)$ in the given range provided $\omega_0 > \omega_r = 3.3$. (This number ω_r was obtained as follows: Let AB be the tangent to the Nyquist locus through the point $(-1/2 + 0j)$; let AC be the line symmetric to AB with respect to the real axis. The intersection of the Nyquist locus and AC then gives $\omega_r/2$.)

This example shows that, although Corollary 6.3.3 did not give an excellent estimate, it is quite simple to apply.

2. Let $G(s) = 1/s(s + 2)$. Determine $K(\omega_0)$ such that the feedback system is L_2-stable for all $k(t) = k(t + T)$, $\omega_0 = 2\pi/T$, and $0 < \epsilon \leqslant k(t) \leqslant K(\omega_0)$. The Nyquist locus of $G(s)$ is shown in Figure 6.7. Using the circle criterion, one obtains $K(\omega_0) = 4$. Brockett (Ref. 11) has shown by examining the worst possible variation in $k(t)$ that $K(\omega_0) = 11.6$. Applying Theorem 6.3 and the graphical procedure outlined in this section results in $K(\omega_0)$ as shown in Figure 6.8. The same figure also shows the result obtained using Corollary 6.3.3 and a graphical construction analogous to the one used in example 1. Thus, by restricting the feedback gain to the periodic, it was possible by means of Theorem 6.3 to obtain higher values of K as the frequency was increased.

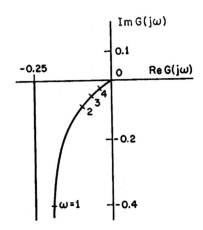

Figure 6.7   Nyquist Locus of $1/s(s + 2)$

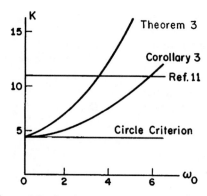

Figure 6.8   Regions of Stability for Example 2

Remark: It follows from example 2 that the converse of Theorems 6.3 is false; i.e., if $F(j\omega)$ does not exist, then there will in general not necessarily be a $k(t)$ in the required range such that the feedback system is not L_2-stable.

Theorem 6.3 has, as is to be expected, an instability converse. It is stated in Theorem 6.4 and can be proved using the methods developed in Chapter 4 and the inequalities introduced in this section.

THEOREM 6.4

Assume that the feedback system by the functional equations (LPFE) is well posed, and that:

1. $\alpha + \epsilon \leqslant k(t) = k(t + T) \leqslant \beta - \epsilon$ for some $\epsilon > 0$ and almost all $t \geqslant T_0$;

2. $\inf_{\mathrm{Re}\, s \geqslant 0} |1 + KG(s)| = 0$ and $\inf_{\mathrm{Re}\, s = 0} |1 + KG(s)| > 0$ for some $K \in [\alpha, \beta]$ where $G(s)$ denotes the Laplace transform of $(g, \{g_n, t_n\})$; and

3. there exists a complex-valued function $F(j\omega) = F(-j\omega)$, of the real variable ω, such that for almost all $\omega \geqslant 0$:

   $F.1.$ $\mathrm{Re}\,\{F(j\omega)\} \geqslant \epsilon$ for some $\epsilon > 0$,

   $F.2.$ $F(j\omega) = F(j(\omega + 2\pi T^{-1})) \in L_\infty$,   and

   $F.3.$ $\mathrm{Re}\left\{F(j\omega)\dfrac{\beta G(j\omega) + 1}{\alpha G(j\omega) + 1}\right\} \geqslant 0.$

Then the feedback system is L_2-unstable.

6.4 A Stability Criterion for Feedback Systems with a Monotone or an Odd-Monotone Nonlinearity in the Feedback Loop

As a second class of stability criteria for feedback systems derived by means of multipliers, consider the system with a time-invariant operator G, in the forward loop and a monotone or an odd-monotone nonlinearity in the feedback loop. For convenience and in order to emphasize the generality of the approaches outlined in Chapters 2 and 4, the results will be derived for systems described by difference equations. With some modifications similar results can be obtained for the continuous case. The adjustments in the theorems involve choosing the integers as the set S and making some minor technical changes. The feedback system which will be considered is shown in Figure 6.9.

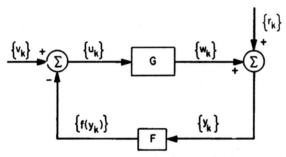

Figure 6.9 The Feedback Loop under Consideration in Section 6.4

Definitions: **The operators G and F are formally defined by**

$$G(\{x_k\})_k = \sum_{l \in I^+} g_{kl} x_l, \qquad k \in I^+,$$

and

$$F(\{x_k\})_k = f(x_k), \qquad k \in I^+,$$

where it is assumed that:

1. $G \in \mathscr{L}^+(l_2, l_2)$, i.e., that G maps l_2 into itself and that $g_{kl} = 0$ whenever $k < l$; and
2. f is a mapping from R into itself for which there exists a k such that $|f(\sigma)| \leqslant K |\sigma|$ for all $\sigma \in R$.

It is simple to verify that under these conditions G and F map l_{2e} into itself and that they are bounded and causal. The system is thus again assumed to be open-loop stable.

The equation describing the forward loop of the feedback system is thus

$$y_k = \sum_{l \in I^+} g_{kl} u_l + r_k, \qquad k \in I^+.$$

The array $\{g_{kl}\}$ is often referred to as the *weighting pattern* of the system. This system is slightly more general than the input-output relation governed by the n-dimensional difference equation

$$x_{k+1} = A_k x_k + b_k u_k,$$
$$y_k = c_k' x_k + d_k u_k, \qquad k \in I^+,$$
$$x_0 = \text{given},$$

where b_k and c_k are n-vectors, d_k is a scalar, A_k is an $(n \times n)$ matrix and x_k is an n-vector called the *state* of the system. This input-output relation is a particular case of the input-output relation defined by the above summation with

$$g_{kl} = \begin{cases} c_k' A_{k-1} \dots A_{l+1} b_l & \text{for} \quad k \geqslant l + 2 \\ c_k' b_k & \text{for} \quad k = l + 1 \\ d_k & \text{for} \quad k = l \\ 0 & \text{otherwise}, \end{cases}$$

$$r_k = c_k' A_{k-1} \dots A_0 x_0 \qquad \text{for} \quad k \geqslant 1,$$

and

$$r_0 = c_0' x_0.$$

The case in which the system is time-invariant is of particular interest. The system is then defined by the equation

$$y_k = \sum_{l \in I^+} g_{k-l} u_l + r_k, \qquad k \in I^+,$$

where g_k is assumed to be zero for $k < 0$. This system is slightly more general than the input-output relation governed by the n-dimensional difference equation

$$x_{k+1} = A x_k + b u_k,$$
$$y_k = c' x_k + d, \qquad k \in I^+,$$

where b and c are constant n-vectors, d is a scalar constant, A is a constant $(n \times n)$ matrix and x_k is an n-vector called the *state* of the system. This input-output relation is a particular case of the input-output relation defined by the above summation with

$$g_k = c'A^{k-1}b \quad \text{for} \quad k > 0,$$

$$g_0 = d$$

$$g_k = 0 \quad \text{for} \quad k < 0,$$

$$r_k = c'A^k x_0 \quad \text{for} \quad k \geqslant 0.$$

The equation describing the feedback loop is

$$u_k = f(y_k) + v_k, \quad k \in I^+,$$

and the closed-loop equation of motion becomes

$$y_k + \sum_{l \in I^+} g_{kl} f(y_l) = \sum_{l \in I^+} g_{kl} v_l + r_k, \quad k \in I^+.$$

It will be *assumed* that this equation is *well posed*. A sufficient condition for well-posedness is that $g_{kk} = 0$ for all $k \in I^+$.

The feedback system under consideration is said to be l_2-*stable* if for all l_2-sequences $r = \{r_k\}$ and $v = \{v_k\}$, all solutions $\{y_k\}$ belong to l_2 and satisfy the inequality

$$\left(\sum_{k \in I^+} y_k^2 \right)^{1/2} \leqslant \rho_1 \left(\sum_{k \in I^+} v_k^2 \right)^{1/2} + \rho_2 \left(\sum_{k \in I^+} r_k^2 \right)^{1/2}$$

for some *constants* ρ_1 and ρ_2.

Remark: Notice that l_2-stability implies that

$$\lim_{k \to \infty} y_k = \lim_{k \to \infty} f(y_k) = 0,$$

and that for the n-dimensional difference equation described above it implies that if $v_k = 0$ for all k then $\lim_{x_0 \to 0} \sup_{k \in I^+} |y_k| = 0$, which in turn implies asymptotic stability in the sense of Lyapunov provided the system is uniformly observable.

Notation and Definitions: The operator F is said to be *monotone* (or *odd-monotone*) if $f(\sigma)$ is a monotone (or an odd-monotone) function of σ. F is said to be *strictly monotone* (or strictly *odd-monotone*) if $f(\sigma) - \epsilon\sigma$ is a monotone (or an odd-monotone) function of σ for some $\epsilon > 0$.

Application of the principles described in the beginning of this chapter and the positive operators introduced in Section 2.5 lead to the following stability theorem.[6] The reader is referred to that section for the nomenclature.

THEOREM 6.5

A sufficient condition for the feedback system under consideration to be l_2-stable is that:

1. G belongs to $\mathscr{L}(l_2,l_2)$ and F is strictly monotone (strictly odd-monotone), and bounded; and
2. there exists an element Z of $\mathscr{L}(l_2,l_2)$, such that $Z - \epsilon I$ is doubly hyperdominant (doubly dominant) for some $\epsilon > 0$ and such that ZG is positive on l_2.

Proof: This theorem is a straightforward application of Corollary 6.2.2 if it can be shown that Z can be factored as required there. This is, however, precisely what is stated in Corollary 3.18.1.

The case in which the system is time invariant and the multiplier is of the Toeplitz type is, of course, of particular interest. The positivity condition and the doubly hyperdominance (doubly dominance) condition can then be stated in terms of z-transforms. This is done in the Corollary 6.5.1.[7]

LEMMA 6.1

Let $R = \{r_{k-l}\}$, $k, l \in I^+$ define an element of $\mathscr{L}(l_2,l_2)$ which is of the Toeplitz type. Then a necessary and sufficient condition for the inner product $\langle x,Rx\rangle$ to be nonnegative for all l_2-summable sequences x is that the z-transform of $\{r_k\}$, $R(z)$, satisfy Re $\{R(z)\} \geqslant 0$ for almost all z with $|z| = 1$.

Proof: It is well known that

$$\langle x,Rx\rangle_{l_2} = \frac{1}{2\pi} \oint_{|z|=1} R(z)\,|X(z)|^2\,z^{-1}\,dz$$

$$= \frac{1}{2\pi} \int_{-\pi}^{\pi} R(e^{j\omega})\,|X(e^{j\omega})|^2\,d\omega$$

$$= \frac{1}{2\pi} \int_{-\pi}^{\pi} \text{Re}\,R(e^{j\omega})\,|X(e^{j\omega})|^2\,d\omega,$$

and the conclusion follows.

---

[6] This result is a generalization of similar results obtained by O'Shea et al. (Refs. 12–14) and Zames and Falb (Ref. 15).
[7] Corollary 6.5.1 is the result of Reference 14.

COROLLARY 6.5.1

A sufficient condition for the feedback system under consideration to be l_2-stable is that:

1. G is a Toeplitz-type element of $\mathscr{L}(l_2,l_2)$, and F is strictly monotone (strictly odd-monotone) and bounded; and
2. there exists a $Z(z)$ such that $Z(z) - \epsilon$ is the z-transform of a hyper-dominant (dominant) sequence for some $\epsilon > 0$ and such that $\mathrm{Re}\,\{G(z)Z(z)\} \geqslant 0$ for almost all z with $|z| = 1$.

Proof: This corollary follows from Theorem 6.5 and Lemma 6.1.

Remark: For the n-dimensional time-invariant difference equation introduced on page 156, the operator G will belong to $\mathscr{L}(l_2,l_2)$ if all eigenvalues of A have magnitude less than unity.

References

1. Popov, V. M., "Absolute Stability of Nonlinear Systems of Automatic Control," *Automation and Remote Control*, Vol. 21, pp. 961–979, 1961.
2. Brockett, R. W., and Willems, J. L., "Frequency Domain Stability Criteria: Part I and II," *IEEE Trans. on Automatic Control*, Vol. AC-10, pp. 255–261 and 401–413, 1965.
3. Zames, G., "On the Input-Output Stability of Time-Varying Nonlinear Feedback Systems. Part I: Conditions Derived Using Concepts of Loop Gain, Conicity, and Positivity; Part II: Conditions Involving Circles in the Frequency Plane and Sector Nonlinearities," *IEEE Trans. on Automatic Control*, Vol. AC-11, pp. 228–238 and 465–476, 1966.
4. Zames, G., "Stability of Systems with Sector Nonlinearities: A Comparison of Various Inequalities," *IEEE Trans. on Automatic Control*, Vol. AC-13, pp. 709–711, 1968.
5. Willems, J. C., *Nonlinear Harmonic Analysis*, Technical Memorandum No. ESL-TM-357, Electronic Systems Laboratory, Massachusetts Institute of Technology, Cambridge, Mass., 1968.
6. Besicovitch, A. S., *Almost Periodic Functions*, Dover Publications, New York, 1956.
7. Bongiorno, J. J., Jr., "Real-Frequency Stability Criteria for Linear Time-Varying Systems," *Proc. IEEE*, Vol. 52, pp. 832–841, 1964.
8. Sandberg, I. W., "On the Stability of Solutions of Linear Differential Equations with Periodic Coefficients," *SIAM J. on Control*, Vol. 12, No. 2, pp. 487–496, 1964.
9. Willems, J. C., "On the Asymptotic Stability of the Null Solution of Linear Differential Equations with Periodic Coefficients," *IEEE Trans. on Automatic Control*, Vol. AC-13, No. 1, pp. 65–72, February 1968.

10. Brockett, R. W., *Finite Dimensional Linear Systems*, John Wiley and Sons, New York, 1970.
11. Brockett, R. W., "Variational Methods for Stability of Periodic Equations," in *Differential Equations and Dynamical Systems* (edited by LaSalle and Hale), pp. 299–308, Academic Press, New York, 1967.
12. O'Shea, R. P., "A Combined Frequency-Time Domain Stability Criterion for Autonomous Continuous Systems," *IEEE Trans. on Automatic Control*, Vol. AC-11, pp. 477–484, 1966.
13. O'Shea, R. P., "An Improved Frequency Time Domain Stability Criterion for Autonomous Continuous Systems," *IEEE Trans. on Automatic Control*, Vol. AC-12, pp. 725–731, 1967.
14. O'Shea, R. P., and Younis, M. I., "A Frequency Time Domain Stability Criterion for Sampled-Data Systems," *IEEE Trans. on Automatic Control*, Vol. AC-12, pp. 719–724, 1967.
15. Zames, G., and Falb, P. L., "Stability Conditions for Systems with Monotone and Slope-Restricted Nonlinearities," *SIAM J. on Control*, Vol. 6, pp. 89–108, 1968.

7 Linearization and Stability

Linear systems are much simpler to design and analyze than nonlinear systems. This is the reason why engineers resort to linear models if this is at all possible and that otherwise a nonlinear system is very often linearized for design purposes. This chapter examines the relationships between stability and continuity of nonlinear systems and their linearizations. It will be shown that a nonlinear system is Lipschitz continuous if and only if its linearization at any point is continuous and that a system whose linearization at the origin is not continuous is not finite-gain stable.

The second part of this chapter contains an account of some of the linearization procedures that are frequently used in engineering design. These include the describing function linearization (often called the equivalent gain or the method of the first harmonic), the total gain linearization, and the incremental gain linearization. An attempt will be made to characterize the philosophy of these methods and to draw attention to their deficiencies. To illustrate this point, the chapter thus ends with a class of counterexamples to Aizerman's conjecture.

7.1 Linearization

The reader is referred to Chapter 2 for the nomenclature and the notation used in this section. Particularly the definitions introduced in Sections 2.4 and 2.6 are freely used.

Definition: Let W_1 and W_2 be Banach spaces and let F be a (in general) nonlinear operator from W_1 into W_2. Let $x_0 \in W_1$ and assume that there exists a bounded linear operator $L_{x_0} \in \mathscr{L}(W_1, W_2)$ such that

$$\lim_{\|x\| \to 0} \frac{\|F(x_0 + x) - Fx_0 - L_{x_0}x\|_{W_2}}{\|x\|_{W_1}} = 0.$$

Then L_{x_0} will be called the *linearization* of F at x_0.

It is in general not clear whether or not a particular operator admits a linearization at a given point. This, as differentiation, indeed requires some smoothness on F. The linearization L_{x_0} is well defined whenever it exists (i.e., it is unique) and preserves some of the properties of F (e.g., its causality).

Some elementary properties of linearizations are:

1. If F admits a linearization at x_0, then F is Lipschitz continuous at x_0.
2. Let F_1 and F_2 be operators from W_1 into W_2 which admit linearizations, L_{1x_0} and L_{2x_0}, at $x_0 \in W_1$ and let α be a scalar. Then $F_1 + F_2$ and αF_1 admit linearizations at x_0. These are, respectively, $L_{1x_0} + L_{2x_0}$ and αL_{1x_0}.
3. Let F_1 and F_2 be operators from, respectively, W_1 into W_2 and W_2 into W_3 which admit linearization L_{1x_0} and $L_{2F_1x_0}$ at, respectively, x_0 and F_1x_0. Then $L_{2F_1x_0}L_{1x_0}$ is the linearization of F_2F_1 at x_0.
4. If $F \in \mathscr{L}(W_1, W_2)$ then L_{x_0} exists for all $x_0 \in W_1$ and $L_{x_0} = F$.

Less elementary but very essential is the following property of linearizations. W_1 and W_2 are from now on assumed to satisfy the axioms which warrant for causality properties and extensions.

THEOREM 7.1

Let F be an operator from W_1 into W_2 and let $x_0 \in W_1$. Assume that F admits the linearization L_{x_0} at x_0. Then L_{x_0} is causal (strongly casual, anticausal, memoryless) on W_1 if F is causal (strongly causal, anti-causal, memoryless) on W_1.

Proof: Let F be causal and assume, to the contrary, that L_{x_0} is not. There then exists an $x \in W_1$ and $T \in S$ (the time-interval of definition) such that $P_T L_{x_0} x \neq 0$ and $P_T x = 0$. Then

$$\lim_{\substack{\alpha \to 0 \\ \alpha \neq 0}} \frac{\|P_T[F(x_0 + \alpha x) - Fx_0 - \alpha L_{x_0}x]\|_{W_2}}{|\alpha| \, \|x\|_{W_1}} = 0,$$

since

$$\|P_T[F(x_0 + \alpha x) - Fx_0 - \alpha L_{x_0}x]\|_{W_2}$$

$$\leqslant \|F(x_0 + \alpha x) - Fx_0 - \alpha L_{x_0}x\|_{W_2}.$$

However,

$$P_T[F(x_0 + \alpha x) - Fx_0] = P_T[FP_T(x_0 + \alpha x) - FP_Tx_0]$$

$$= P_T(FP_Tx_0 - FP_Tx_0)$$

$$= 0$$

implies that

$$\lim_{\substack{\alpha \to 0 \\ \alpha \neq 0}} \frac{\|P_T L_{x_0}x\|_{W_2}}{\|x\|_{W_1}} = \frac{\|P_T L_{x_0}x\|_{W_2}}{\|x\|_{W_1}},$$

which yields the desired contradiction. The other cases are proved in a similar fashion.

One of the main points of this monograph is the importance of causality of operators and the resulting possibility of analyzing systems on extended spaces. The question thus arises whether or not the definition of linearization can be extended to cover causal operators on extended spaces. This is indeed possible.

Definition: Let W_1 and W_2 be Banach spaces defined on the time-interval of definition S and let W_{1e} and W_{2e} denote their extensions (it is thus assumed that W_{1e} and W_{2e} satisfy the axioms of extended spaces). Let F be a causal operator from W_{1e} into W_{2e}. Then the operator L_{x_0} from W_{1e} into W_{2e} is said to be a *linearization* of F at x_0 if it is linear and if for all $T \in S$, $P_T L_{x_0}$ is a linearization of $P_T F$ at $P_T x_0$. It is easily verified that L_{x_0} exists for any causal operator which has a linearization at $P_T x_0$ for all $T \in S$. This follows from Theorem 7.1 and by *defining* $P_T L_{x_0}$ to be the linearization of $P_T F$ at $P_T x_0$. The resulting linearized operator L_{x_0} is then defined on the extended space by $(P_T L_{x_0}x)(t) = (L_{x_0,T}x)(t)$ for $t \leqslant T$, $t \in S$, with $L_{x_0,T}$ the linearization of $P_T F$ at $P_T x_0$. Clearly this provides an alternative characterization of L_{x_0}. Linearizations on extended spaces are again well defined whenever they exist and satisfy the elementary properties 1–4. The following theorem is now obvious.

THEOREM 7.2

Let F be a causal operator from W_{1e} into W_{2e} and let $x_0 \in W_{1e}$. Assume that F admits a linearization L_{x_0} at x_0. Then L_{x_0} is causal (strongly causal, memoryless) on W_{1e} if F is causal (strongly causal, memoryless) on W_{1e}.

The following examples serve to convince the reader that the functional linearization considered here specializes to the more familiar concepts of linearization as encountered in the theory of functions on Euclidean spaces and in the theory of ordinary differential equations.

1. Let $f \colon R^n \times S \to R^m$ be differentiable for all $\sigma \in R^n$ and for almost all $t \in S$ and assume that for some $K < \infty$, $\|f(\sigma_1,t) - f(\sigma_2,t)\| \leqslant K \|\sigma_1 - \sigma_2\|$ for all $\sigma_1, \sigma_2 \in R^n$ and almost all $t \in S$. Let $(Fx)(t) \triangleq f(x(t),t)$. It is clear that F is a well-defined memoryless operator from $L_{pe}^{R^n}(S)$ into $L_{pe}^{R^m}(S)$ for $1 \leqslant p \leqslant \infty$ and that F admits a linearization for all $x_0 \in L_{pe}^{R^n}$. In fact, $(F_{x_0}x)(t) = \frac{\partial f}{\partial \sigma}(x_0(t),t)x(t)$, where $\frac{\partial f}{\partial \sigma}(\sigma,t)$ is the ($m \times m$) Jacobian matrix (the (i,j)th entry of $\partial f/\partial \sigma$ equals $\partial f_i(\sigma,t)/\partial \sigma_j$).

2. Consider the nth order ordinary differential equation

$$\dot{x} = f(x,u,t),$$

$$y = g(x,u,t),$$

with $S = [T_0,\infty)$, $u \in R^m$, $x \in R^n$, $y \in R^k$, and $x(t_0) \in R^n$ given. Assume appropriate smoothness conditions on f and g (e.g., f, g uniformly Lipschitz and differentiable). Let F be defined by $y \triangleq Fu$, where u generates y through the differential equation. Then F maps $L_{pe}^{R^m}(S)$ into $L_{pe}^{R^k}(S)$, $1 \leqslant p \leqslant \infty$, is causal on $L_{pe}^{R^m}(S)$, and F admits a linearization for any $u_0 \in L_{pe}^{R^m}(S)$ with L_{u_0} defined through the *linear* ordinary differential equation

$$\Delta \dot{x} = \frac{\partial f}{\partial x}(x_0,u_0,t)\,\Delta x + \frac{\partial f}{\partial u}(x_0,u_0,t)u,$$

$$y = \frac{\partial g}{\partial x}(x_0,u_0,t)\,\Delta x + \frac{\partial g}{\partial u}(x_0,u_0,t)u,$$

with $\Delta x(t_0) = 0$, u_0 given, and x_0 the corresponding solution.

The above examples demonstrate that time invariance of operators is *not* preserved under linearization. If, however, the element $x_0 \in W_{1e}$ is itself constant for all $t \in S$, if $S = (-\infty,+\infty)$, and if F is time invariant, then it follows readily that L_{x_0} will also be time invariant whenever it exists.

Theorem 7.3

Let $F \in \tilde{\mathscr{B}}(W,W)$ be regular in $\tilde{\mathscr{B}}(W,W)$ and assume that F has a linearization L_{x_0} at $x_0 \in W$. Then $L_{x_0}^{-1}$ exists and is the linearization of F^{-1} at Fx_0.

Proof that L_{x_0} is one-to-one: Assume to the contrary that $L_{x_0}x = 0$ for some $x \in W$, $x \neq 0$. Then

$$\lim_{\substack{\alpha \to 0 \\ \alpha \neq 0}} \frac{\|F(x_0 + \alpha x) - Fx_0\|}{\|\alpha x\|} = 0,$$

which shows that

$$\lim_{\substack{\alpha \to 0 \\ \alpha \neq 0}} \frac{\|F(x_0 + \alpha x) - Fx_0\|}{\|F^{-1}(F(x_0 + \alpha x)) - F^{-1}Fx_0\|} = 0,$$

and that F^{-1} is not Lipschitz continuous at x_0.

Proof that L_{x_0} is onto: Let F_0 be defined as $F_0 x = F(x + x_0) - Fx_0$ and let F_0^{-1} be its inverse. This inverse exists since F is by assumption invertible. Consider now the equation $x = Hx \triangleq x - L_{x_0}F_0^{-1}x + y$. It follows from the definition of a linearization that

$$\frac{\|x_1 - x_2 - L_{x_0}(F_0^{-1}x_1 - F_0^{-1}x_2)\|}{\|F_0^{-1}x_1 - F_0^{-1}x_2\|} \to 0 \quad \text{as} \quad \|x_1 - x_2\| \to 0.$$

In particular since F_0^{-1} is Lipschitz continuous,

$$\|x_1 - x_2 - L_{x_0}(F_0^{-1}x_1 - F_0^{-1}x_2)\| < K \|x_1 - x_2\|$$

with $K < 1$ for $\|x_1 - x_2\|$ sufficiently small. This shows that H is a contraction on some sphere around the origin. A similar calculation yields that for y sufficient small H maps this sphere into itself[1] and thus has a fixed point. Hence the equation $L_{x_0}F_{x_0}^{-1}x = y$ has a solution for y sufficiently small which by linearity of L_{x_0} shows that L_{x_0} is indeed onto as claimed.

Proof that $L_{x_0}^{-1}$ is bounded and that it linearizes F^{-1} at Fx_0: The inverse $L_{x_0}^{-1}$ is bounded by the closed graph theorem. To see that $L_{x_0}^{-1}$ linearizes F^{-1} at Fx_0 note that since

$$\|F^{-1}(Fx_0 + L_{x_0}x) - F^{-1}Fx_0 - x\|$$
$$\leqslant \|F^{-1}\| \, \|Fx_0 + L_{x_0}x - F(x_0 + x)\|,$$

[1] The method of proof suggested here follows the usual proofs of the implicit function theorem (see, e.g., Ref. 1, p. 47).

it follows that

$$\lim_{\|x\| \to 0} \frac{\|F^{-1}(Fx_0 + L_{x_0}x) - F^{-1}Fx_0 - x\|}{\|x\|} = 0,$$

which shows that

$$\lim_{\|L_{x_0}^{-1}z\| \to 0} \frac{\|F^{-1}(Fx_0 + z) - F^{-1}Fx_0 - L_{x_0}^{-1}z\|}{\|L_{x_0}^{-1}z\|} = 0.$$

Since $\|L_{x_0}^{-1}z\| \geqslant \|L_{x_0}\|^{-1} \|z\|$, it follows that

$$\lim_{\|z\| \to 0} \frac{\|F^{-1}(Fx_0 + z) - F^{-1}Fx_0 - L_{x_0}^{-1}z\|}{\|z\|} = 0.$$

Theorem 7.3 and the definition of linearization on extended spaces combine to give

THEOREM 7.4

Let F be a causal locally Lipschitz continuous operator from W_e into itself with a causal locally Lipschitz continuous inverse on W_e. Assume that F admits the linearization L_{x_0} at x_0. Then L_{x_0} is invertible, L_{x_0} and $L_{x_0}^{-1}$ are causal and locally Lipschitz continuous on W_e, and F^{-1} admits the linearization $L_{x_0}^{-1}$ at Fx_0.

7.2. Linearization, Stability, and Continuity

In this section the fundamental relationships between stability and continuity of a nonlinear feedback system and of its inverse are considered in detail. First, however, a fundamental lemma is proven.

LEMMA 7.1

Let F be an operator from W_1 into W_2 and assume that F admits a linearization at every $x_0 \in W_1$. Let L_{x_0} denote this linearization. Then $\|L_{x_0}\|$ is uniformly bounded in $\mathscr{L}(W_1, W_2)$ if and only if F is Lipschitz continuous on W_1. In fact, $F\|_\Delta = \sup_{x_0 \in W_1} \|L_{x_0}\|$.

Proof: It will be shown that $\|F\|_\Delta = \sup_{x_0 \in W_1} \|L_{x_0}\|$, where the equality means that if one side exists, so does the other, and they are equal. The inequality $\|L_{x_0}\| \leqslant \|F\|_\Delta$ follows immediately from the definition of linearization and yields $\sup_{x_0 \in W_1} \|L_{x_0}\| \leqslant \|F\|_\Delta$. To prove the converse inequality, let $x, y \in W_1$ be given and consider the equality

$$Fx - Fy = \int_0^1 \frac{dF}{d\alpha} (x - \alpha(y - x)) \, d\alpha = \int_0^1 L_{x-\alpha(y-x)}(x - y) \, d\alpha.$$

This yields $\|Fx - Fy\| \leqslant \sup_{\alpha \in [0,1]} \|L_{x-\alpha(y-x)}\| \|x - y\|$, and thus

$$\|F\|_{\Delta} \triangleq \sup_{x,y \in W_1} \frac{\|Fx - Fy\|_{W_2}}{\|x - y\|_{W_1}} \leqslant \sup_{x_0 \in W_1} \|L_{x_0}\|.$$

Hence $\sup_{x_0 \in W_1} \|L_{x_0}\| = \|F\|_{\Delta}$ as claimed.

Consider now the feedback system studied in Chapter 4 that is described by the functional equations

$$e_1 = u_1 - y_2,$$
$$e_2 = u_2 + y_1,$$
$$y_1 = G_1 e_1, \qquad \qquad \text{(FE)}$$
$$y_2 = G_2 e_2,$$

and satisfies the conditions W.1–W.4, G.1–G.4, and I.1 enumerated in Chapter 4. The time-interval of definition is $S = [T_0, \infty)$, and well-posedness imposes weak conditions on the operators G_1 and G_2. One such condition has been given in Theorem 4.1.

Assume that for some elements $x_1 \in W_{1e}$ and $x_2 \in W_{2e}$ the operators G_1 and G_2 admit linearizations L_{1x_1} and L_{2x_2} at, respectively, x_1 and x_2. Consider now the functional equations associated with this linearization given by

$$e'_1 = u'_1 - y'_2,$$
$$e'_2 = u'_2 + y'_1,$$
$$y'_1 = L_{1x_1} e'_1, \qquad \qquad \text{(LFE)}$$
$$y'_2 = L_{2x_2} e'_2.$$

It follows immediately from Theorem 7.2 that these equations satisfy the assumptions W.1–W.4, G.1–G.4, and I.1 if $u'_i \in W_{ie}$, $i = 1, 2$. This system obviously describes a feedback system that will be called the *linearized feedback system*. Note that this linearized feedback system is linear but that the operators in the forward and the feedback loop depend on the point of linearization, $(x_1, x_2) \in W_{1e} \times W_{2e}$.

Theorem 7.5 exposes the relationship between continuity of a feedback system and its linearization.[2]

---

[2] Theorem 7.5 is an infinite-dimensional version of a theorem of Palais (Refs. 2, 3) which essentially states that a nonlinear transformation on a finite-dimensional space is invertible if all its linearizations are invertible. Theorem 7.5 is by no means a generalization of this theorem to infinite-dimensional spaces, since it exploits causality and invertibility on extended spaces in a very essential manner.

THEOREM 7.5

Consider the feedback system described by the functional equations (FE) and assume that it is well posed. Assume that the operator G_i, $i = 1, 2$, admits a linearization at every $x_i \in W_{ie}$. Then the (in general nonlinear) feedback system described by the functional equations (FE) is Lipschitz continuous if and only if the linearized feedback system described by the functional equations (LFE) is Lipschitz continuous for all $(x_1, x_2) \in W_{1e} \times W_{2e}$, uniformly in (x_1, x_2).

Proof: It can be shown that well-posedness of (FE) implies well-posedness of (LFE). However, since the thrust of Theorem 7.5 lies in the continuity implications and since the demonstration of the above claim is rather involved, its proof will be deleted. Turning now to the continuity proof, let G denote the operator defined on $W_{1e} \times W_{2e}$ by $G(e_1, e_2) = (-G_2 e_2, G_1 e_1)$. Clearly, $L(e_1, e_2) = (-L_{x_2} e_2, L_{x_1} e_1)$ is the linearization of G at (x_1, x_2). By Theorem 7.4, $(I + G)^{-1}$ (which exists and is causal by well-posedness) admits the linearization $(I + L)^{-1}$ at (x_1, x_2). Thus by Lemma 7.1,

$$\|P_T(I + G)^{-1} P_T\|_\Delta = \sup_{(x_1, x_2) \in W_{1e} \times W_{2e}} \|P_T(I + L)^{-1} P_T\| \quad \text{for all} \quad T \in S.$$

Now passing to the limit with $T \to \infty$ shows that these limits exist simultaneously — since both sides are monotone in T — and that they are equal if they exist. This is precisely the claim of the theorem.

Theorem 7.5 thus shows that for Lipschitz continuity it suffices to consider the linearized system. The following theorem relates stability of the nonlinear system with the stability of its linearization at the origin. The resulting theorem is consequently a great deal weaker.

THEOREM 7.6

Consider the feedback system described by the functional equations (FE) and assume that it is well posed. Assume that the operator G_i, $i = 1, 2$, admits a linearization at $0 \in W_{ie}$. Then the (in general nonlinear) feedback system described by the functional equations (FE) is not finite-gain stable (and thus not Lipschitz continuous) if the feedback system linearized at the origin (i.e., described by the functional equations (LFE) with $(x_1, x_2) = 0$) is not continuous.

Proof: Let G denote the operator as defined in the proof of Theorem 7.5 and let L_0 be its linearization at the origin. Let $u = (u_1, u_2) \in W_1 \times W_2$ be the input to both the linearized and the nonlinear feedback system. Assume now that the nonlinear feedback system is finite-gain stable; i.e., $\|(I + G)^{-1}\| < \infty$. Since $(I + L_0)^{-1}$ is the linearization of $(I + G)^{-1}$ at the origin, it follows that for all $T \in S$,

$$
\begin{aligned}
0 = \lim_{\alpha \to 0} & \frac{\|P_T[(I + G)^{-1}\alpha u - (I + L_0)^{-1}\alpha u]\|}{|\alpha| \, \|u\|} \\
\geqslant \lim_{\alpha \to 0} & \left(\frac{\|P_T(I + L_0)^{-1}u\|}{\|u\|} - \frac{\|P_T(I + G)^{-1}\alpha u\|}{|\alpha| \, \|u\|} \right) \\
\geqslant & \frac{\|P_T(I + L_0)^{-1}u\|}{\|u\|} - \|(I + G)^{-1}\|.
\end{aligned}
$$

Hence $\|P_T(I + L_0)^{-1}u\| \leqslant \|(I + G)^{-1}\| \, \|u\|$, which yields stability of the linearized system and the proof of the theorem.

The ideas of Theorems 7.5 and 7.6 can be combined to show that a well-posed feedback system is continuous if and only if it is stable and its linearization at any point is continuous. This thus relaxes the uniformity requirement of Theorem 7.5, but introduces the explicit requirement that the feedback system be stable. Since this latter condition cannot be stated in terms of linearizations, the combined theorems cannot provide a test for verifying continuity by solely considering linearizations, and so the combination has not been stated explicitly in this section.

7.3 The Describing Function, the Total Gain Linearization, and the Incremental Gain Linearization

In the previous chapters, a number of stability criteria for nonlinear feedback systems have been derived. The question of whether or not these criteria are conservative cannot be given a general answer, but both from the estimates and from examples one suspects that these criteria are by no means necessary *and* sufficient (Ref. 4). Thus the question arises whether these criteria are inadequate or too conservative, and if instability and stability conditions can be derived using some approximate methods.

There have indeed been a number of such attempts in the engineering literature. The best known of these approximate methods are the describing function (and its generalizations), the total gain linearization based on Aizerman's conjecture, and the incremental gain linearization. All these methods rest in essence on a common technical (but not mathematical) principle. In particular, the describing function has been very extensively — and, it appears, successfully — applied to engineering problems.

There is one class of feedback systems for which necessary and sufficient conditions for stability are known, namely, the Nyquist criterion, which applies to feedback systems with a linear, time invariant system in the forward loop and a constant feedback gain. Thus, by associating with a nonlinear feedback system a well-chosen class of linear time invariant feedback systems, one tries to conclude stability or instability.

The following section takes a critical look at some of these linearization procedures and exposes, by means of an example, unexpected periodic solutions in a class of nonlinear feedback systems. Although the system chosen to obtain this conclusion might seem quite special, the method of analysis remains applicable to other systems and will expose an essentially similar behavior. The example could also suggest to what extent and for which systems the existing frequency-domain stability criteria can be improved. They also demonstrate the need for caution in applying these linearization techniques in stability analysis.

Consider the feedback system shown in Figure 7.1. The relation

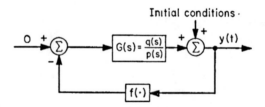

Figure 7.1   Feedback System

between the input and the output of the element in the forward loop is determined by the ordinary time-invariant differential equation

$$\dot{x}(t) = Ax(t) + bu(t),$$

$$y(t) = c'x(t),$$

where A is a constant ($n \times n$) matrix, and b and c are constant n-vectors. The transfer function of this system is thus given by $G(s) = c'(Is - A)^{-1}b$ and is the ratio of two polynomials in s with the degree of the numerator less than the degree of the denominator. The element $f(\cdot)$ in the feedback loop generates an output $f(\sigma)$ when its input is σ, and f is a Lipschitz-continuous mapping from the real line into itself. The differential equation describing the closed-loop system is thus

$$\dot{x}(t) = Ax(t) - bf(c'x(t)).$$

It is assumed that $f(0) = 0$. The solution $x(t) \equiv 0$ is called the *null solution* of this system and is said to be *asymptotically stable in the large* if it is stable (in the sense of Lyapunov) and if all solutions converge to the null solution for $t \to \infty$. For convenience the feedback system under consideration is said to be asymptotically stable in the large if the null solution is.

Rather than investigating input-output stability as in the remainder of this monograph, attention is focused in this section on asymptotic stability, in order that the reader may more easily compare the literature on these linearization methods. It can actually be shown, using the methods employed in Section 5.3, that L_p-stability, with $1 \leqslant p \leqslant \infty$, of the feedback system under consideration implies asymptotic stability in the large.

For the case for which $f(\sigma) = K\sigma$, this stability problem can be completely resolved using root-locus techniques, the Nyquist stability criterion or a Routh-Hurwitz test (Ref. 5) and thus presents in principle no difficulties. If $f(\sigma)$ is nonlinear, however, this is not so, and often in engineering practice the question whether a particular feedback system of this type is asymptotically stable in the large is answered by considering a linearized model.

Three common types of linearization are the dc type of linearization, the ac type of linearization, and the describing-function type of linearization. These are now formally defined.

Definitions: Let f be a mapping from the real line into itself with $f(0) = 0$. The dc *gain* or the *total gain* of the nonlinearity $f(\sigma)$ at σ ($\sigma \neq 0$) is defined by $K_t(\sigma) = f(\sigma)/\sigma$. If f is differentiable, then the ac *gain* or the *incremental gain* of the nonlinearity $f(\sigma)$ at σ is defined by $K_i(\sigma) = \partial f(\sigma)/\partial \sigma$. If f satisfies for some M_1, M_2 the inequality $|f(\sigma)| \leqslant M_1 + M_2 |\sigma|$ for all σ, then the *describing-function gain* (or the *equivalent gain*) of the nonlinearity $f(\sigma)$ at amplitude $A (A \neq 0)$ is the complex number $K_d(A)$ defined by

$$K_d(A) = \frac{1}{\pi A} \int_0^{2\pi} f(A \cos t) e^{jt} \, dt.$$

Clearly, $K_d(A) = (1/\pi A) \int_0^{2\pi} f(A \cos t) \cos t \, dt$, where f is a single-valued function. The describing function can also be used when f is a hysteresis loop or a similar multiple-valued memory-type nonlinearity. In that case the equivalent gain would in general be a complex number. One also often requires the nonlinearity to be odd; this is an attempt to eliminate the dc term in the output with a sinusoidal input. It should also be noted that the describing-function method has been extended so that it allows for more than simple sinusoidal inputs and that it is capable of treating uncertainty as well.[3]

The procedure by which linearization attempts to conclude stability for the dc, the ac, and the equivalent-gain types of linearization goes as follows: If the linear system with $f(\sigma) = K\sigma$ is asymptotically stable for all K in the range of the dc gain, the ac gain, or the equivalent gain (i.e., for all $K = K_t(\sigma)$, $K = K_i(\sigma)$, or $K = K_d(A)$ and all σ or A) then the nonlinear system is asymptotically stable in the large.

Both the dc type and the ac type of linearization and the resulting conclusions about stability have been the subject of rather well-known conjectures, due, respectively, to Aizerman (Ref. 7) and Kalman (Ref. 8). Particularly the Aizerman conjecture has received a lot of attention.[4]

The intuitive reasoning behind these conjectures in rather clear: in the total gain linearization (Aizerman's conjecture), one realizes that asymptotic stability is a property pertaining to the origin, and thus one replaces the nonlinearity by a linear gain that, viewed from the origin, gives instantaneously the same gain. It is clear that if stability results for all *time-varying* gains in the range of the dc gain, then the feedback system is indeed asymptotically stable in the large. However, there does not seem to be any reason to expect that the above procedure yields correct results. Indeed, it sometimes fails. For incremental gain linearization, one views the local properties of the nonlinearity and replaces the nonlinearity by a linear gain which is locally the same. One

---

[3] An excellent source for the describing function and its generalizations is Reference 6.

[4] Originally published in 1949, it took until 1958 before Pliss (Ref. 9) gave a satisfactory counterexample. It is possible to show using the Popov criterion that for second-order systems the conjecture is true with the exception of some cases where the dc gain approaches for large values of its argument a gain for which the resulting linear system is not asymptotically stable. The counterexample given by Krasovskii (Ref. 10) is in fact of this kind. The counterexamples obtained by Pliss, however, are more satisfactory. The very stringent conditions on the nonlinearity and the involved mathematics kept the work of Pliss from being well known. More recently, Dewey and Jury (Ref. 11) and Fitts (Ref. 12) gave numerical counterexamples derived from a computer simulation. The conjecture due to Kalman in which the ac gain is used predicts stability in the large only for a subclass of the nonlinearities for which Aizerman's conjecture does. Fitts (Ref. 12) gives counterexamples to this conjecture derived from a computer analysis.

ignores in this process the fact that this linear gain does not extend to the origin. The incremental gain does correspond to a type of linearization: namely those linearizations obtained by considering constant inputs. The linearization thus suggests a continuity consideration, but the conclusion pertains to stability. Note that the range of the dc gain is always *smaller* than the range of the ac gain and thus that the ac gain type of linearization leads to more cautious conclusions than the dc gain type of linearization.

The question of continuity of a particular feedback system can be answered by linearization, as indicated in Section 7.2. However, one needs thereby to replace the nonlinearity f by a gain $k(t)$, which varies between $\alpha \leqslant k(t) \leqslant \beta$ with $\alpha = \inf_{\sigma \in R} K_i(\sigma)$ and $\beta = \sup_{\sigma \in R} K_i(\sigma)$, and the resulting linear system needs to be stable for any gain satisfying the constraint $\alpha \leqslant k(t) \leqslant \beta$.

The describing-function technique is clearly inspired by linear time-invariant systems where the eigenfunctions (in L_∞) are sinusoids, and thus the responses to sinusoids are completely representative of the system behavior. This, however, does not carry over to nonlinear systems and makes the mathematical philosophy of the describing function unclear. However, it appears to yield good results, which is understandable in view of the fact that if a system sustains oscillations, then it is very likely that the first harmonics are in balance somewhere in the neighborhood of the oscillation.

In what follows, a simple, rigorous proof of the existence of periodic solutions in a fourth-order system will be given. It will be shown, however, that all of the above-mentioned linearization techniques predict asymptotic stability in the large. These oscillations thus constitute a class of counterexamples to both Aizerman's conjecture and Kalman's suggestion. The results are obtained using perturbation theory (Refs. 13, 14). Since the ideas behind this technique are simple, the theorem from which the results follow will be proved.

7.4 Averaging Theory

Consider the ordinary differential equation

$$\dot{x}(t) = Ax(t) + \epsilon f(x(t), z, \epsilon),$$

where $x(t)$ is an element of R_n, A is a constant $(n \times n)$ matrix, z is a parameter (an element of R_m), ϵ is a scalar parameter, and f is a mapping from $R_n \times R_m \times R$ into R_n such that for all R, ϵ_0, and M there exists a constant $K(R,\epsilon_0,M)$ (the Lipschitz constant) such that $\|f(x_1,z,\epsilon) - f(x_2,z,\epsilon)\| \leqslant K \|x_1 - x_2\|$ for all $\|x_1\|, \|x_2\| \leqslant R, |\epsilon| \leqslant \epsilon_0$ and $\|z\| \leqslant M$.

Since the function f does not satisfy a global Lipschitz condition, it is not clear at this point whether a solution $x(t)$ to the differential equation exists for all t. This problem is resolved in the next lemma.

Definition: Let $x(t)$ be a continuous map from $[0,T]$ into a Banach space. Then $\sup_{t \in [0,T]} \|x(t)\|$ exists and is called the *norm induced by the uniform topology*. This space is complete. Recall that the contraction mapping principle states that if F is a map from a complete metric space X into itself with $d(F(x),F(y)) \leqslant \alpha\, d(x,y)$ for all x, $y \in X$ and some $\alpha < 1$, then the equation $x = Fx$ has a unique solution (called a *fixed point* of the mapping F). Moreover, picking any x_0 and defining $x_{k+1} \triangleq Fx_k, k \in I^+$, yields a sequence $\{x_k\}$ which converges in the metric of X to the fixed point of F.

LEMMA 7.2

Given any $\tau > 0$, ρ, and M, then the above differential equation has a unique solution $x(t)$ for any $x(0), z,$ and t that satisfy $\|x(0)\| \leqslant \rho$, $0 \leqslant t \leqslant \tau$, and $\|z\| \leqslant M$ provided ϵ is sufficiently small (i.e., for all ϵ with $|\epsilon| \leqslant \epsilon_1$ and some $\epsilon_1 > 0$). Moreover, this solution can be obtained using the successive approximations

$$x_0(t) = e^{At}x(0)$$

and

$$x_{k+1}(t) = e^{At}x(0) + \epsilon \int_0^t e^{A(t-\sigma)} f(x_k(\sigma),z,\epsilon)\, d\sigma$$

for $k \in I^+$.

Proof: Let S be the sphere in the Banach space of all continuous mappings from $[0,\tau]$ into R_n with the uniform topology and with $\|x(t)\| \leqslant 2\rho N$ where $N = \sup_{0 \leqslant t \leqslant \tau} \|e^{At}\|$. The mapping F defined on S by

$$Fx(t) = e^{At}x(0) + \epsilon \int_0^t e^{A(t-\sigma)} f(x(\sigma),z,\epsilon)\, d\sigma,$$

maps S into itself for all $|\epsilon| \leqslant \epsilon_1$ with

$$\epsilon_1 \leqslant \min \{\epsilon_0, (KN_\tau)^{-1}, \rho\tau^{-1}(4\rho NK + \|f(0,0,0)\|)^{-1}\},$$

where K is the Lipschitz constant associated with $R = 2\rho N$, $\epsilon_0 > 0$, and M. Moreover, F is a contraction on S. The verification of these facts is simple and will not be given explicitly. Thus, the equation $x(t) = Fx(t)$ has a unique fixed point, which can be obtained using the successive approximations as stated. This yields the lemma.

The next lemma exposes the dependence of $x(t)$ on ϵ more explicitly:

LEMMA 7.3

Given any $\tau > 0$, ρ, and M, then the solution $x(t)$ to the above differential equation for any $x(0)$, z, and t that satisfy $\|x(0)\| \leqslant \rho, 0 \leqslant t \leqslant \tau$, and $\|z\| \leqslant M$ can be expressed as

$$x(t) = e^{At}x(0) + \epsilon \int_0^t e^{A(t-\sigma)}f(e^{A\sigma}x(0),z,\epsilon)\,d\sigma + \epsilon^2 L(t,x(0),z,\epsilon)$$

for all ϵ sufficiently small (i.e., for all ϵ with $|\epsilon| \leqslant \epsilon_2$ and some $\epsilon_2 > 0$). Moreover, $L(t,x(0),z,\epsilon)$ is bounded for $0 \leqslant t \leqslant \tau$, $\|x(0)\| \leqslant \rho$, $\|z\| \leqslant M$, and $|\epsilon| \leqslant \epsilon_2$.

Proof: It will be shown that the $(k + 1)$th element in the series of successive approximations introduced above is of this form provided the kth one is, and that the bound on L_k can be taken to be independent of k. Since $x_1(t)$ is clearly of that form, the result follows by induction, because the limit for $k \to \infty$ exists and must also be of this form. Let K be the Lipschitz constant associated with $2\rho N$, ϵ_1, and M, and let $\epsilon_2 < \min\{\epsilon_1,(2N\tau)^{-1}\}$. A simple calculation then shows that $\|L_{k+1}\| \leqslant \tau^2 N^2(\|f(0,0,0)\| + KN\rho)$ if $\|L_k\| \leqslant \tau^2 N^2(\|f(0,0,0)\| + KN\rho)$, which then, in view of the above remarks, yields the lemma.

Lemma 7.3 yields the following theorem on the existence of periodic solutions to the differential equation under consideration.

THEOREM 7.7

If for ϵ sufficiently small (i.e., for all ϵ with $|\epsilon| \leqslant \epsilon_0$ and some $\epsilon_0 > 0$) there exist bounded functions $x(0)(\epsilon)$, $T(\epsilon)$, and $z(\epsilon)$ such that

$$x(0)(\epsilon) = e^{AT(\epsilon)}x(0)(\epsilon) + \epsilon \int_0^{T(\epsilon)} e^{A[T(\epsilon)-\sigma]}f(e^{AT(\epsilon)}x(0)(\epsilon),z(\epsilon),\epsilon)\,d\sigma$$
$$+ \epsilon^2 L(T(\epsilon),x(0)(\epsilon),z(\epsilon),\epsilon),$$

then the differential equation under consideration has a periodic solution for ϵ sufficiently small (i.e., for ϵ with $|\epsilon| \leqslant \epsilon_1$ and some $\epsilon_1 > 0$).

Proof: Lemma 7.2 shows that for ϵ sufficiently small $x(T(\epsilon)) = x(0)(\epsilon)$, which, since the differential equation under consideration is time-invariant, yields a periodic solution of period $T(\epsilon)$.

Theorem 7.7 is not very useful as it stands, since it requires computation of the function L and solving for the functions $x(0)(\epsilon)$, $T(\epsilon)$ and $z(\epsilon)$. However, by using the implicit function theorem it is possible to obtain sufficient conditions for the conditions of Theorem 7.7 to be satisfied.

In the theorem which follows, use will be made of the *implicit function theorem* (Ref. 1, p. 47), which states that if f maps $R_n \times R_m$ into R_n and if:

1. $f(x_0, y_0) = 0$ for some $x_0 \in R_n$, $y_0 \in R_m$,
2. $\frac{\partial f}{\partial y}(x, y)$ exists and is continuous in a neighborhood of the point x_0, y_0, and
3. $\frac{\partial f}{\partial y}(x_0, y_0)$ is of rank n,

then there exists a map, ϕ, from R_n into R_m, which is continuous in a neighborhood of x_0 and such that $y = \phi(x)$ yields $f(\phi(x), x) = 0$ for all x in some neighborhood of x_0. Moreover, $y_0 = \phi(x_0)$, and ϕ is unique in a neighborhood of x_0.

THEOREM 7.8

Assume that $e^{AT_0} = I$ (i.e., that all solutions of $\dot{x}(t) = Ax(t)$ are periodic with period T_0), and that $f(x, z, \epsilon)$ is a continuous function of x, z, and ϵ that has continuous first partial derivatives with respect to x and z for ϵ sufficiently small (i.e., for all ϵ with $|\epsilon| \leqslant \epsilon_0$ and some $\epsilon_0 > 0$.) Let

$$F(x, z, \epsilon) \triangleq \int_0^{T_0} e^{-A\sigma} f(e^{A\sigma} x, z, \epsilon)\, d\sigma$$

and assume: 1, that $F(x_0, z_0, 0) = 0$; and 2, that the matrix $\frac{\partial F}{\partial x, \partial z}(x_0, z_0, 0)$ is of full rank. Then there exists a continuous function $z(\epsilon)$ such that for ϵ sufficiently small (i.e., for all ϵ with $|\epsilon| \leqslant \epsilon_1$ and some $\epsilon_1 > 0$) the differential equation under consideration has a periodic solution $x^*(t, \epsilon)$ with $\lim_{\epsilon \to 0} z(\epsilon) = z_0$ and $\lim_{\epsilon \to 0} x^*(t, \epsilon) = e^{At} x_0$.

Proof: The smoothness conditions on f together with the resulting smoothness of the solutions of ordinary differential equations (Ref. 15, p. 29) ensure that the implicit function theorem is applicable. This in turn shows that the conditions 1 and 2 of the theorem are sufficient to ensure that Theorem 7.7 is applicable, which leads to the conclusion of the theorem.

This method of concluding the existence of periodic solutions for differential equations is known as *Averaging Theory* since the function F as defined in Theorem 7.8 requires the average value of the velocity vector along the solution $e^{At}x_0$ to be zero.

7.5 Counterexamples to Aizerman's Conjecture

Consider the differential equation

$$x^{(4)}(t) + 10x^{(2)}(t) + 9x(t) + \epsilon[\alpha x^{(3)}(t) + \beta x^{(2)}(t)$$
$$+ \gamma x^{(1)}(t) + \delta x(t)] + \epsilon f(x^{(2)}(t)) = 0,$$

where f maps R into R and is continuously differentiable with respect

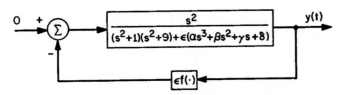

Figure 7.2   The Fourth-Order System to which Averaging Theory is Applied and which Yields Counterexamples to Aizerman's Conjecture

to its argument. This equation describes the feedback system shown in Figure 7.2 and is equivalent to the following system of first-order differential equations:

$$
\begin{bmatrix} \dot{z}_1(t) \\ \dot{z}_2(t) \\ \dot{z}_3(t) \\ \dot{z}_4(t) \end{bmatrix} =
\begin{bmatrix} 0 & 1 & 0 & 0 \\ -1 & 0 & 0 & 0 \\ 0 & 0 & 0 & 3 \\ 0 & 0 & -3 & 0 \end{bmatrix}
\begin{bmatrix} z_1(t) \\ z_2(t) \\ z_3(t) \\ z_4(t) \end{bmatrix}
$$

$$
+ \frac{\epsilon}{8} \left(\begin{bmatrix} 0 \\ -1 \\ 0 \\ \frac{1}{3} \end{bmatrix} \begin{bmatrix} \delta - \beta & \gamma - \alpha & \delta - 9\beta & 3(\gamma - 9\alpha) \end{bmatrix} \begin{bmatrix} z_1(t) \\ z_2(t) \\ z_3(t) \\ z_4(t) \end{bmatrix} + \begin{bmatrix} 0 \\ 1 \\ 0 \\ -3 \end{bmatrix} f(z_1(t) + z_3(t)) \right) + O(\epsilon^2),
$$

where $O(\epsilon^2)$ denotes a four-dimensional vector which is such that $\lim_{\epsilon \to 0} [O(\epsilon^2)/\epsilon] = 0$. The application of Theorem 7.8 shows that there exist continuous functions $\alpha(\epsilon)$, $\beta(\epsilon)$, $\gamma(\epsilon)$, and $\delta(\epsilon)$ such that the differential equation under consideration has a periodic solution, $z^*(t,\epsilon)$, with $\lim_{\epsilon \to 0} \alpha(\epsilon)$, $\beta(\epsilon)$, $\gamma(\epsilon)$, $\delta(\epsilon) = \alpha_0, \beta_0, \gamma_0, \delta_0$ and

$$\lim_{\epsilon \to 0} z^*(t,\epsilon) = e^{At} \begin{bmatrix} z_{1,0} \\ z_{2,0} \\ z_{3,0} \\ z_{4,0} \end{bmatrix} \qquad A = \begin{bmatrix} 0 & 1 & 0 & 0 \\ -1 & 0 & 0 & 0 \\ 0 & 0 & 0 & 3 \\ 0 & 0 & -3 & 0 \end{bmatrix}$$

if

(1) $(\gamma_0 - \alpha_0)z_{1,0} + (\beta_0 - \delta_0)z_{2,0} + \dfrac{1}{\pi} \displaystyle\int_0^{2\pi} f(z_{1,0} \cos \sigma$

$$+ z_{2,0} \sin \sigma + z_{3,0} \cos 3\sigma + z_{4,0} \sin 3\sigma) \sin \sigma \, d\sigma = 0,$$

$(\beta_0 - \delta_0)z_{1,0} - (\gamma_0 - \alpha_0)z_{2,0} + \dfrac{1}{\pi} \displaystyle\int_0^{2\pi} f(z_{1,0} \cos \sigma$

$$+ z_{2,0} \sin \sigma + z_{3,0} \cos 3\sigma + z_{4,0} \sin 3\sigma) \cos \sigma \, d\sigma = 0,$$

$(\gamma_0/3 - 3\alpha_0)z_{3,0} + (\beta_0 - \delta_0/9)z_{4,0} + \dfrac{1}{\pi} \displaystyle\int_0^{2\pi} f(z_{1,0} \cos \sigma$

$$+ z_{2,0} \sin \sigma + z_{3,0} \cos 3\sigma + z_{4,0} \sin 3\sigma) \sin 3\sigma \, d\sigma = 0,$$

$(\beta_0 - \delta_0/9)z_{3,0} - (\gamma_0/3 - 3\alpha_0)z_{4,0} + \dfrac{1}{\pi} \displaystyle\int_0^{2\pi} f(z_{1,0} \cos \sigma$

$$+ z_{2,0} \sin \sigma + z_{3,0} \cos 3\sigma + z_{4,0} \sin 3\sigma) \cos 3\sigma \, d\sigma = 0,$$

and

(2) $(z_{1,0}^2 + z_{2,0}^2)(z_{3,0}^2 + z_{4,0}^2) \neq 0.$

Equation (2) guarantees that the matrix in Theorem 7.8 is of full rank, and equation (1) is derived from the requirement that the average be zero.

From these conditions the following theorem, which will be central in establishing the counterexamples to Aizerman's conjecture is now derived.

THEOREM 7.9

If $f(\sigma)$ is not identically equal to $k\sigma$ for any constant k, then there exists a nonzero periodic solution to the differential equation under consideration for ϵ sufficiently small (i.e., for all ϵ with $|\epsilon| \leqslant \epsilon_0$ and $\epsilon_0 > 0$), and proper choice of the functions $\alpha(\epsilon)$, $\beta(\epsilon)$, $\gamma(\epsilon)$, and $\delta(\epsilon)$. Moreover, the functions $\alpha(\epsilon)$ and $\gamma(\epsilon)$ which yield this periodic solution satisfy the inequality

$$(\gamma(\epsilon) - \alpha(\epsilon))(\gamma(\epsilon) - 9\alpha(\epsilon)) < 0.$$

Proof: The proof proceeds as follows: First, α_0, β_0, γ_0, and δ_0 will be chosen in such a way that the determining equations are satisfied. If $z_{1,0}$, $z_{2,0}$, $z_{3,0}$, and $z_{4,0}$ are chosen such that $(z_{1,0}^2 + z_{2,0}^2)(z_{3,0}^2 + z_{4,0}^2) \neq 0$, then the determining equations can be solved uniquely for α_0, β_0, γ_0, and δ_0, since the determinant of these linear equations in α_0, β_0, γ_0, and δ_0 is then nonzero and yields in particular the following expression for $\gamma_0 - \alpha_0$:

$$\gamma_0 - \alpha_0 = \frac{1}{z_{1,0}^2 + z_{2,0}^2} \frac{1}{\pi} \int_0^{2\pi} f(z_{1,0} \cos \sigma + z_{2,0} \sin \sigma$$

$$+ z_{3,0} \cos 3\sigma + z_{4,0} \sin 3\sigma)(z_{1,0} \sin \sigma - z_{2,0} \cos \sigma) \, d\sigma.$$

If f is linear, then $\gamma_0 - \alpha_0 = 0$. If f is not linear, then there exists a choice of $z_{1,0}$, $z_{2,0}$, $z_{3,0}$, and $z_{4,0}$ with $(z_{1,0}^2 + z_{2,0}^2)(z_{3,0}^2 + z_{4,0}^2) \neq 0$ such that $\gamma_0 - \alpha_0 \neq 0$. Moreover, from these determining equations it also follows that

$$(\gamma_0 - \alpha_0)(z_{1,0}^2 + z_{2,0}^2) + (\gamma_0 - 9\alpha_0)(z_{3,0}^2 + z_{4,0}^2) = 0.$$

Thus $(\gamma_0 - \alpha_0)(\gamma_0 - 9\alpha_0) < 0$ for this choice of $z_{1,0}$, $z_{2,0}$, $z_{3,0}$, and $z_{4,0}$ with $(z_{1,0}^2 + z_{2,0}^2)(z_{3,0}^2 + z_{4,0}^2) \neq 0$, which yields the conclusion of the theorem.

Consider now the zeros of the polynomial

$$s^4 + 10s^2 + 9 + \epsilon(\alpha s^3 + \beta s^2 + \gamma s + \delta) + \epsilon K s^2.$$

For ϵ sufficiently small and for K bounded, these zeros lie:

1. in Re $s < 0$ if $\epsilon > 0$, $\alpha > 0$, $\gamma > 0$, and $(\gamma - \alpha)(\gamma - 9\alpha) < 0$,

    or if $\epsilon < 0$, $\alpha < 0$, $\gamma < 0$, and $(\gamma - \alpha)(\gamma - 9\alpha) < 0$;

2. in Re $s > 0$ if $\epsilon < 0$, $\alpha > 0$, $\gamma > 0$, and $(\gamma - \alpha)(\gamma - 9\alpha) < 0$,

    or if $\epsilon > 0$, $\alpha < 0$, $\gamma < 0$, and $(\gamma - \alpha)(\gamma - 9\alpha) < 0$.

Thus all the linearization techniques would predict that the feedback system under consideration is asymptotically stable in the large provided that $\epsilon > 0$, $\alpha > 0$, $\gamma > 0$ and $(\gamma - \alpha)(\gamma - 9\alpha) < 0$ or that $\epsilon < 0$, $\alpha < 0$, $\gamma < 0$ and $(\gamma - \alpha)(\gamma - 9\alpha) < 0$ and the linearized gain satisfies $0 \leqslant K \leqslant 1$. These regions are graphically shown in Figure 7.3.

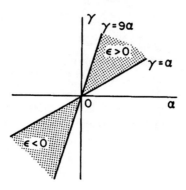

Figure 7.3   Conditions on ϵ, α, γ to Obtain Counterexamples to Aizerman's Conjecture

Choosing now a nonlinearity such that $K_d(\sigma)$, $K_i(\sigma)$, and $K_d(A)$ satisfy $0 \leqslant K \leqslant 1$ (e.g., $f(\sigma) = \tanh \sigma$), it is clear that for ϵ sufficiently small and for values of α and γ such that $(\gamma - \alpha)(\gamma - 9\alpha) < 0$, the sign of ϵ can be chosen in such a way that the linearization techniques would predict the feedback system under consideration to asymptotically stable in the large. This, however, is in direct contradiction with Theorem 7.9, which shows that the feedback system sustains a periodic solution.

Remark 1: The choice of the function $f(\sigma) = \tanh \sigma$, is irrelevant. In fact, the same conclusion holds for any nonlinearity, provided it is sufficiently smooth for Theorem 7.9 to be applicable and provided $|f(\sigma)| \leqslant K |\sigma|$ for some K and all σ which then yields, for ϵ sufficiently small, the pole locations of the linearized system as given.

Remark 2: The remarkable feature of the periodic solutions discovered in Theorem 7.9 is that (for ϵ sufficiently small) they occur only when the linearized system has all its poles either always in the left half-plane or always in the right half-plane, contrary to what is to be expected from linearization.

Remark 3: The Nyquist locus and the root locus of the fourth-order system under consideration are shown in Figure 7.4 for the case

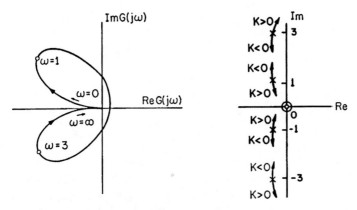

Figure 7.4 Nyquist Locus of $G(j\omega)$ and Root Locus of the Linearized Feedback System

$\epsilon > 0$, $\alpha > 0$, $\gamma > 0$, and $(\gamma - \alpha)(\gamma - 9\alpha) > 0$ or $\epsilon < 0$, $\alpha < 0$, $\gamma < 0$, and $(\gamma - \alpha)(\gamma - 9\alpha) < 0$.

Remark 4: The local stability properties of these periodic solutions is of course of interest. Variational techniques show that for proper choices of α, β, γ, σ, ϵ, and $f(\cdot)$ these periodic solutions can be locally stable.

The existence of the periodic solutions discovered in this section will now be given an intuitive explanation.[5] This will of course be a plausibility argument. Averaging theory allows us to conclude that the argument is correct provided ϵ is sufficiently small.

Assume an input to the nonlinearity $\epsilon f(\cdot)$ which has a first harmonic, a third harmonic, and "small" other harmonics. The output to the nonlinearity will thus contain all harmonics, all of comparable magnitudes, and all "small" since they have been multiplied by a small parameter ϵ. Let x_1, x_3, y_1, and y_3 be the Fourier coefficients of the first and the third harmonics of the input and the output to the nonlinearity. It can be shown that for particular choices of x_1 and y_1 the nonlinearity will shift the phases of the first and third harmonics toward one another,

---

[5] The dual-input describing function would predict these oscillations. However, the method of proof will yield counterexamples to the dual-input describing function when applied to higher-order systems.

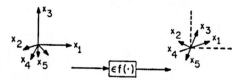

Figure 7.5 The Frequency Spectrum of the Input and the Output of the Element in the Feedback Loop

thus obtaining the situation depicted in Figure 7.5. The negative feedback leads to an input u to the forward loop — as is shown in Figure 7.6 — that with a Nyquist locus as in Figure 7.6 multiplies the first and third harmonic by a factor of order ϵ^{-1} and shifts their phases in the right direction but by an amount less than 180°, thus restoring the original relationship of x_1 and x_3. The higher harmonics remain of order ϵ. The loop can thus be closed and the feedback system sustains the oscillation.

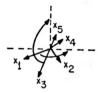

Figure 7.6 The Frequency Spectrum of the Input and the Output of the Element in the Forward Loop

References

1. Saaty, T. L., *Modern Nonlinear Equations*, McGraw-Hill, New York, 1967.
2. Palais, R. C., "Natural Operations on Differential Forms," *Trans. Amer. Math. Soc.*, Vol. 92, No. 1, pp. 125–141, 1959.
3. Holzmann, C. A., and Liu, R., "On the Dynamical Equations of Nonlinear Networks with n-Coupled Elements," *Proceedings of the Third Annual Allerton Conference on Circuit and System Theory*, pp. 536–545, 1965.
4. Brockett, R. W., "Variational Methods for Stability of Periodic Equations," in *Differential Equations and Dynamical Systems*, (edited by LaSalle and Hale), pp. 299–308, Academic Press, New York, 1967.
5. Newton, G. C., Jr., Gould, L. A., and Kaiser, J. F., *Analytical Design of Linear Feedback Controls*, John Wiley and Sons, New York, 1957.

6. Gelb, A., and Vander Velde, W. E., *Multiple-Input Describing Functions and Nonlinear System Design*, McGraw-Hill, New York, 1968.

7. Aizerman, M. A., "On a Problem Concerning the Stability in the Large of Dynamical Systems," *Uspekhi Mat. Nauk*, Vol. 4, pp. 187–188, 1949.

8. Kalman, R. E., "Physical and Mathematical Mechanisms of Instability in Nonlinear Automatic Control Systems," *Trans. ASME*, Vol. 79, pp. 553–566, 1957.

9. Pliss, V. A., *Certain Problems in the Theory of the Stability of Motion in the Whole*, Leningrad University Press, Leningrad, 1958.

10. Krasovskii, N. N., "On the Stability in the Whole of the Solutions of a System of Nonlinear Differential Equations," *Prikl. Mat. Mekh.*, Vol. 18, pp. 735–737, 1954.

11. Dewey, A. G., and Jury, E. T., "A Note on Aizerman's Conjecture," *IEEE Trans. on Automatic Control*, Vol. AC-10, pp. 482–483, 1965.

12. Fitts, R. E., "Two Counterexamples to Aizerman's Conjecture," *IEEE Trans. on Automatic Control*, Vol. AC-11, pp. 553–556, 1966.

13. Cesari, L., "Existence Theorems for Periodic Solutions of Nonlinear Lipschitzian Differential Equations," in *Contributions to the Theory of Nonlinear Oscillations*, Vol. 5, Princeton University Press, Princeton, N.J., 1960.

14. Hale, J. K., *Oscillations in Nonlinear Systems*, McGraw-Hill, New York, 1963.

15. Coddington, E. A., and Levinson, N., *Theory of Ordinary Differential Equations*, McGraw-Hill, New York, 1955.

Index

185

Unbiased sequences, 58, 63

Vector space, 3

W, W, notation, 13
Well-posedness of feedback systems,
  conditions, 96
  definition, 90
  discussion. 93
  examples, 95
Wiener-Hopf equation, 80

z-transform, 7